Collins

AQA GCSE 9-1
Food Preparation and Nutrition

Revision Guide

Kath Callaghan, Fiona Balding,
Barbara Monks, Barbara Rathmill
and Suzanne Gray with Louise T. Davies

About this Revision & Practice book

Revise

These pages provide a recap of everything you need to know for each topic.

You should read through all the information before taking the Quick Test at the end. This will test whether you can recall the key facts.

Quick Test
1. What are the fat-soluble vitamins?
2. Give **three** good sources of vitamin C.
3. Which vitamin is connected with the healthy development of the spine in unborn babies?

Practise

These topic-based questions appear shortly after the revision pages for each topic and will test whether you have understood the topic. If you get any of the questions wrong, make sure you read the correct answer carefully.

Review

These topic-based questions appear later in the book, allowing you to revisit the topic and test how well you have remembered the information. If you get any of the questions wrong, make sure you read the correct answer carefully.

Mix it Up

These pages feature a mix of questions for all the different topics, just like you would get in an exam. They will make sure you can recall the relevant information to answer a question without being told which topic it relates to.

Test Yourself on the Go

Visit our website at **collins.co.uk/collinsGCSErevision** and print off a set of flashcards. These pocket-sized cards feature questions and answers so that you can test yourself on all the key facts anytime and anywhere. You will also find lots more information about the advantages of spaced practice and how to plan for it.

Workbook

This section features even more topic-based questions as well as practice exam papers, providing two further practice opportunities for each topic to guarantee the best results.

ebook

To access the ebook revision guide visit

collins.co.uk/ebooks

and follow the step-by-step instructions.

Contents

	Revise	Practise	Review

Food Preparation Skills

	Revise	Practise	Review
Knife Skills	p. 6 ☐	p. 16 ☐	p. 32 ☐
Fish	p. 8 ☐	p. 16 ☐	p. 32 ☐
Meat	p. 10 ☐	p. 17 ☐	p. 33 ☐
Prepare, Combine and Shape	p. 12 ☐	p. 18 ☐	p. 34 ☐
Dough	p. 14 ☐	p. 19 ☐	p. 35 ☐

Food Nutrition and Health

	Revise	Practise	Review
Protein and Fat	p. 20 ☐	p. 36 ☐	p. 50 ☐
Carbohydrate	p. 22 ☐	p. 36 ☐	p. 50 ☐
Vitamins	p. 24 ☐	p. 37 ☐	p. 51 ☐
Minerals and Water	p. 26 ☐	p. 38 ☐	p. 51 ☐
Making Informed Choices	p. 28 ☐	p. 38 ☐	p. 52 ☐
Diet, Nutrition and Health	p. 30 ☐	p. 39 ☐	p. 52 ☐

Food Science

	Revise	Practise	Review
Cooking of Food, Heat Transfer and Selecting Appropriate Cooking Methods	p. 40 ☐	p. 54 ☐	p. 68 ☐
Proteins and Enzymic Browning	p. 42 ☐	p. 54 ☐	p. 68 ☐
Carbohydrates	p. 44 ☐	p. 56 ☐	p. 69 ☐
Fats and Oils	p. 46 ☐	p. 56 ☐	p. 70 ☐
Raising Agents	p. 48 ☐	p. 57 ☐	p. 72 ☐

Contents

	Revise	Practise	Review

Food Safety

	Revise	Practise	Review
Microorganisms, Enzymes and Food Spoilage	p. 58 ☐	p. 73 ☐	p. 86 ☐
Microorganisms in Food Production	p. 60 ☐	p. 73 ☐	p. 86 ☐
Bacterial Contamination	p. 62 ☐	p. 74 ☐	p. 87 ☐
Buying and Storing Food	p. 64 ☐	p. 74 ☐	p. 88 ☐
Preparing and Cooking Food	p. 66 ☐	p. 75 ☐	p. 89 ☐

Food Choices

	Revise	Practise	Review
Food Choices	p. 76 ☐	p. 90 ☐	p. 104 ☐
British and International Cuisines	p. 78 ☐	p. 90 ☐	p. 104 ☐
Sensory Evaluation	p. 80 ☐	p. 91 ☐	p. 105 ☐
Food Labelling	p. 82 ☐	p. 92 ☐	p. 106 ☐
Factors Affecting Food Choices	p. 84 ☐	p. 93 ☐	p. 107 ☐

Food Provenance

	Revise	Practise	Review
Food and the Environment	p. 94 ☐	p. 108 ☐	p. 112 ☐
Food Provenance and Production Methods	p. 96 ☐	p. 108 ☐	p. 112 ☐
Sustainability of Food	p. 98 ☐	p. 109 ☐	p. 113 ☐
Food Production	p. 100 ☐	p. 110 ☐	p. 114 ☐
Food Processing	p. 102 ☐	p. 110 ☐	p. 114 ☐

Mixed Questions	p. 116		
Answers	p. 132		
Glossary	p. 140		
Index	p. 144		

Knife Skills

You must be able to:

- Understand two different methods of using knives to prepare food safely
- Explain the techniques used when preparing different foods that require knife skills.

Knife Holds

Claw Grip

- To use the **claw grip**, shape your hand into a claw shape, tucking the thumb inside the fingers – the knuckle to fingertips part of the hand acts as a barrier against the knife blade when being held in the claw grip shape. It is safer to use a large knife with a flat-sided blade than a smaller one for this reason.
- Place the item you want to cut flat side down on a chopping board and rest the claw on the item to be sliced.
- Hold the knife in the other hand. Use the knife point as a pivot (it should not leave the board). As you slice, the food moves towards the knife; this reduces the health and safety risk.

Claw grip

Bridge Hold

- To use the **bridge hold**, first place the flat surface of the item on a chopping board.
- Now form a bridge with the thumb and index finger of one hand and hold the item on the chopping board.
- Hold a knife in the other hand and position the blade under the bridge formed with your hand. Firmly cut downwards.

Bridge hold

Knife Safety Rules

- The correct knife should be used for the appropriate job.
- Knives must be kept sharp and clean; a blunt knife is more likely to cause a cut because more pressure needs to be applied to use it to cut.
- Knife handles must be grease-free.
- The point must always be downwards when carrying a knife.
- Knives should not be put in the washing-up bowl.
- A knife must not be left on the edge of a table or chopping board.

Knife Skills for Vegetable Preparation

- The classic cuts for vegetables are shown in the table.

 Key Point

Specific types of knives are designed for different cutting and shaping tasks.

 Key Point

Knives are dangerous if not handled correctly and care should be taken at all times.

A flat and stable cutting surface is essential to avoid injury when cutting food.

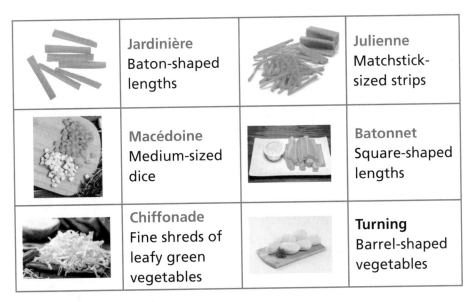

	Jardinière Baton-shaped lengths		**Julienne** Matchstick-sized strips
	Macédoine Medium-sized dice		**Batonnet** Square-shaped lengths
	Chiffonade Fine shreds of leafy green vegetables		**Turning** Barrel-shaped vegetables

> **Key Point**
>
> There are specific terms used for vegetable cuts relating to the size and shape of the outcome.

Specific Types of Knives

Knife	Description	Uses
Cook's knife	Comes in different sizes. Strong, ridged blade is suitable for a range of tasks.	Dicing, chopping and trimming vegetables, meat, poultry, and fresh herbs.
Paring knife	A small knife with a thin and slightly flexible blade.	Fruit and vegetable preparation.
Boning knife	A very strong blade that will not bend or break easily. May have a straight or curved blade.	Removing bones from meat joints and poultry.
Filleting knife	Thin-bladed, flexible, very sharp knife.	Filleting fish.
Carving knife	Long blade with a serrated or plain edge. Can be rounded or pointed.	Carving meat joints or cooked hams.
Bread knife	Long serrated edge.	Slicing loaves and other bread products.
Palette knife	Flexible blade, which is rounded at the top.	Icing cakes; turning food during cooking; moulding and smoothing food.

> **Key Words**
>
> claw grip   dicing
> bridge   chopping
> hold   paring
> jardinière   flexible
> julienne   filleting
> macédoine   serrated
> chiffonade   cooking
> batonnet

> **Quick Test**
>
> 1. How would you carry a knife?
> 2. Is a blunt knife safer to use than a sharp knife?
> 3. Why is a flat surface needed to prepare food?
> 4. Name the two methods of holding food when cutting it.

Fish

You must be able to:

- Know how to classify different types of fish
- Explain how to choose, handle and prepare different types of fish.

Classifications of Fish

- There are three main classifications of fish:
  - **White fish** has flesh that is white in colour and contains less fat. White fish can be round or flat, e.g. cod, haddock.
  - **Flat fish** – examples include plaice, sole and halibut.
  - **Oily fish** has flesh that is coloured and contains more fat than white fish but this is healthy fat with fish oils. Examples include salmon, sardines, mackerel and tuna.
- There are two main classifications of **shellfish**:
  - **Crustaceans** – crabs, lobsters, prawns, crayfish, shrimp and squid (this has a hard backbone that must be removed during preparation).
  - **Molluscs** – oysters, mussels, scallops, winkles and cockles.

Nutrients Found in Fish

- Nutrients in fish include:
  - Protein – for growth and repair.
  - Minerals – iron, zinc and iodine – for red blood cells, metabolism and to regulate blood sugar.
  - Vitamins A and D – for vision, body linings, the immune system, bone health, and to help with mineral absorption.
- Oily fish contains Omega 3 fatty acids for brain development, healthy bones and joints.

> ### Key Point
>
> White fish carry oil in the liver; oily fish carry oil throughout the flesh.

Buying Quality Fish

- To ensure you are buying quality fish, you need to make sure that: eyes are bright, not dull; scales are in place; gills are bright red; it has a slightly salty, fresh smell of the sea (fish smells bad as it deteriorates); it has a thin layer of slime; the flesh is firm.
- Shellfish must be prepared and eaten within two days of purchase.

Preserving Fish

- There are a number of ways of preserving fish.
- **Canning** – cans are heated to 121 °C to kill bacteria; this heating creates a vacuum inside the can. Any blown or dented cans can cause food poisoning as the vacuum is broken and clostridium botulinum food poisoning can occur.

- **Freezing** – sea fish are frozen within 90 minutes of capture to a minimum temperature of –18 °C. This stops food poisoning bacteria from reproducing and forces the bacteria to become dormant – it does not kill the bacteria.
- **Smoking** – to 76 °C or above removes moisture from the fish and gives a distinctive flavour, e.g. smoked salmon.
- **Salting** – salt is added to fish to remove its moisture. Food poisoning bacteria cannot survive without moisture.

Key Point

It's important to wash your hands after handling fish to prevent cross-contamination.

Preparing and Cooking Raw Fish

To Prepare	To Cook
<ul><li>Use a blue chopping board (sanitized).</li><li>Use a filleting knife (sharp and sanitized).</li><li>Check for freshness.</li></ul>	<ul><li>Fish cooks quickly because the muscle is short and the **connective tissue** is thin. The connective tissue is made up of collagen and will change into gelatine and **coagulate** at 75 °C.</li><li>Fish can be grilled, baked or fried. Often fish is **enrobed** in breadcrumbs/batter to protect it when using high heat.</li><li>Fish can also be cooked gently by steaming or poaching without coating the flesh.</li></ul>

Filleting a Flat Fish

	Use a filleting knife to descale and remove the fins. Cut off the head just behind the gills.
	Cut from head to tail down to the bone, to one side of the centre line.
	Turn the knife almost parallel to the table. Make a long, smooth cut. Cut horizontally against the backbone towards the outer edge. Separate the fillet from the bone and remove it.

Filleting a Round Fish

	Descale and remove the fins. Cut into the top of the fish on one side of the tail; detach the backbone from head to tail.
	Cut under the flesh towards the tail and detach the cut piece.
	Cut along the curved rib bones and finish detaching the fillet at the head. Turn the fish over and repeat to remove the second fillet.

Quick Test

1. Which type of fish contains the most Omega 3 fatty acids?
2. Describe **two** quality checks for fresh fish.
3. What equipment is needed when preparing raw fish?

Key Words

crustacean	connective
mollusc	tissue
smoking	coagulate
salting	enrobed

Meat

You must be able to:

- Know the main sources of meat and how they are prepared
- Understand the structure of meat and how this affects the cooking method used.

Nutritional Content

- Meat contains:
  - Protein, including **collagen**, **elastin** and **myoglobin**, which makes the meat red in colour.
  - Fat (saturated) – provides warmth and protection of the animal's internal organs.
  - Minerals, e.g. iron, calcium and phosphorus – needed to form red blood cells, bones and teeth, and for energy metabolism.
  - Vitamins B6 and B12 – needed to release energy from foods.
  - Cholesterol.

> **Key Point**
>
> The length and type of cooking method depends on the type of **muscle fibre**.

Classifications of Meat

- Meat is classified as the muscle tissue of dead animals and birds.
- There are four main meat sources:
  - Animals: pork, (pigs), beef (cattle), lamb (sheep).
  - Poultry: chicken, turkey, duck, goose.
  - Game: feathered or furred; venison, rabbit, pheasant.
  - Offal: liver, tongue, tripe, kidney, heart, brain, trotters.

The Structure of Meat

- Meat is a muscle made of cells, which consist of fibres held together by connective tissue. Long fibres are associated with tough meat – the older an animal is, the tougher the meat.
- Muscles that work a lot, such as the thighs and shoulders of animals, give tough meat, e.g. shin beef, brisket. Small fibres are associated with tender cuts.

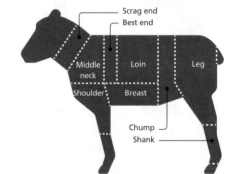

Meat Cuts and Cooking

- Cuts of meat from muscle areas that do a lot of work will need longer, slower cooking methods in wet heat, e.g. stewing, braising, pot roasting and casseroling.
- Meat from tougher cuts can be ground or minced to break up the connective tissues so that it cooks more quickly.
- Cuts of meat from muscle areas not so heavily used by the animal, e.g. the back and the rump, can be cooked much more quickly in dry heat, e.g. grilling, stir-frying.

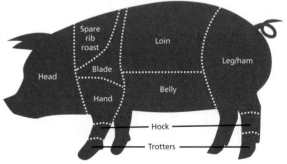

Marinating and Tenderising

- Marinades are added to meat before cooking to add flavour and the acid content (e.g. lemon juice, yoghurt, wine) breaks down the protein.
- Meat is tenderised by: using a marinade; mincing; or using a steak hammer.

The Effects of Cooking Meat

- The browning of meat is caused by a reaction with natural sugars and proteins to produce a dark colour. This is called the **Maillard reaction** or **non-enzymic browning**.
- As meat cooks, the proteins coagulate due to heat. Collagen breaks down into **gelatine**, making the meat tender.

Cooking Different Meat Joints

Meat	Temperature	Time per 500 g
Pork	75–80 °C	30 mins + 15 mins
Poultry	75–80 °C	20 mins + 20 mins
Beef	Rare: 52 °C	20 mins
Lamb	Well done: 75–80 °C	30 mins

Checking for Readiness

- You should know the safety rules for cooking different meat.
- Meat joints can be tested using a meat probe or temperature probe. To determine if a steak is rare, medium or well done the 'poke' test can be learned and used.
- The following foods should not be eaten if undercooked: chicken (80 °C), pork (75 °C), offal and game, burgers, sausages, kebabs.

Boning a Chicken

- Do not wash the chicken. Place **whole chicken** on a red board.
- Remove the legs. Cut down through the skin and between the joints. Turn the chicken over and break the legs.
- Find the knuckle and cut through the leg to separate the **drumstick** and the **thigh**.
- Cut through the joints to remove the **wings**.
- Find the wishbone at the front of the bird. Cut a V-shape on either side of the bird to release the wishbone, then cut through the knuckle at the base. Carefully remove the **breasts** from the carcass.

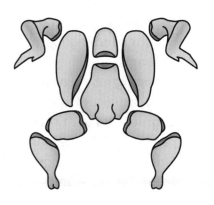

Storing Meat Safely

- Raw meat should be kept separate from cooked meat to avoid **cross-contamination**.
- Raw meat should be stored covered at the bottom of the fridge so meat juices cannot drip onto other foods.
- Cooked meat should be kept chilled and covered and treated as a high-risk food.
- Chilled meat should be stored at between 1 °C and 4 °C.
- Frozen meat should be stored at between –18 °C and –22 °C.

> ### Quick Test
>
> 1. Where would you store meat when not preparing it?
> 2. Tough meat has what length of fibres?
> 3. What are the main sources of meat?

> ### Key Words
>
> collagen
> elastin
> myoglobin
> muscle fibre
> Maillard reaction
> non-enzymic browning
> gelatine
> cross-contamination

Prepare, Combine and Shape

You must be able to:

- Understand that a recipe consists of specific quantities of ingredients that are prepared, using a variety of skills, to produce the required outcome
- Know that ingredients can be combined and shaped to make the finished dish more attractive and appetising to the consumer.

Methods for Cutting and Preparing Ingredients

- **Slicing:** slicing even-sized cuts to suit the requirements of the dish.
- **Peeling:** using a peeler or knife to remove skin from fruit or vegetables, although it can be beneficial to leave skin on as it contains Non-Starch Polysaccharides (NSP) or fibre.
- **Chopping:** cutting **ingredients** into small pieces, **precise** sizes or roughly chopped.
- **Dicing:** cutting into cubes.
- **Grating:** passing food over a blade to reduce food in order to achieve the required size or thickness.
- **Coring:** removing the centre core of vegetables and fruit.
- **Mashing/crushing:** pressing into a puree or crushing into small pieces perhaps with the skin left on for added fibre.
- **Shredding:** grating or slicing into thin long flexible strips.
- **Scissor snipping:** using scissors to cut to the required shape, size or length.
- **Scooping:** using a spoon, scoop or melon baller to remove small circular pieces from the ingredient.
- **Segmenting:** cutting even-sized pieces from the ingredient, removing pith, zest and membrane.
- **Skinning:** peeling, grating, zesting or slicing off the outer skin from the ingredient.
- **Blanching:** placing the ingredient into boiling water for a few minutes to quickly cook, soften and inhibit the enzymic activity.
- **Blending:** using an electric food processor or blender to process different food ingredients together to form one product or sauce.
- **Juicing:** squeezing an ingredient to remove the liquid from it.
- **Preparing garnishes** – for example, zesting, carving, making ribbons, crisps, twists, roses, and using edible flowers.

Peeling food

Grating food

Combining and Shaping Ingredients

- Ingredients can be **combined** or mixed in a number of ways:
  - Whisking – combining ingredients to incorporate air into a mixture, e.g. meringues.
  - Stirring – gently mixing ingredients with a spoon.
  - Folding – gently incorporating mixtures together, such as flour into a cake mixture.

> **Key Point**
>
> Enzymic activity occurs when cut fruit and vegetables react with oxygen to turn them brown.

Stirring food (Stirfry)

- Creaming – mixing two ingredients together, such as sugar and butter in cake making.
- **Rubbing-in** – incorporating fat into flour when making pastry.
- Ingredients can be shaped in a number of ways:
  - By hand – to achieve an even shape, for example, when making fishcakes, meatballs and koftas.
  - In a mould – e.g. to form the shape of bread loaves, mousses.
  - Using cutters – dough can be cut precisely to keep consistency.
  - With a rolling pin – used to achieve a desired thickness.
  - Using a piping bag – shaped nozzles create patterns, e.g. duchess potato.

Binding, Coating and Glazing

Binding

- **Binding** means holding ingredients together.
- Egg acts as a binding agent to hold ingredients together in burgers.
- Water binds flour and fat in pastry to help form a dough; some enriched shortcrust pastries use egg yolk to bind.
- Potato and/or flour are used as a binding material in fish cakes.
- Breadcrumbs are used as a binder in sausage mixtures.

Coating

- **Coating** means adding an outer layer to a food.
- Breadcrumbs are used for coating fish cakes and chicken **goujons**.
- Batters are used to enrobe, coat and protect fish and fritters that are deep-fried.
- Chocolate can be used as a coating to hold food together.

Glazing

- Egg wash **glaze** has a golden, shiny finish and is good for pastry and dough.
- Egg white has a crisp, golden texture and is good for sweet pastry dishes.
- Egg yolk gives a golden brown colour and works well on potato dishes.
- Milk gives a golden brown glaze for scones, pastry and biscuits.
- Sugar and water form a sweet sticky covering used on **enriched dough**.
- Jam gives a shiny finish and is brushed over fruit flans.
- Arrowroot is a clear shiny gel that is used to finish fruit flans.

 Key Point

Various foods can be coated with ingredients to create a new layer to protect, add texture and flavour – this is called coating, or enrobing.

 Key Words

ingredients
precise
combined
rubbing-in
binding
coating
goujons
enriched dough

Quick Test

1. How would you shape meatballs?
2. What would you use to shape a burger?
3. What glaze would you use on enriched dough?

Dough

You must be able to:

- Know that making and shaping dough is a precursor to making a variety of flour-based mixtures
- Understand the function of ingredients in dough.

Bread Dough

- Bread dough is made with a strong plain flour, which contains a high level (17%) of the proteins **gliadin** and **glutenin**.
- **Gluten** is produced when water is added to the flour, enabling strong elastic dough to be formed.
- The elasticity of the dough helps to trap **carbon dioxide gas** (CO_2), which is produced by the yeast to raise the dough.

The Function of Ingredients in Bread

- Each ingredient in bread has a specific function:
  - A **strong flour** is used as it has a high gluten content.
  - **Yeast** is the raising agent, producing carbon dioxide.
  - **Warm liquid** is needed to form the dough and encourage the yeast to grow.
  - **Salt** adds flavour, and aids gluten development.
  - **Fat** extends the shelf-life and adds colour and flavour.

Key Point

Dough is made by mixing flour with liquid, and sometimes includes leavening (raising) agents as well as other ingredients and flavourings.

Enriched Dough

- **Enriched dough** is a basic bread dough with the addition of sugar, butter and sometimes egg.
- Additional flavourings such as mixed spices, dried fruit and nuts can be added.
- Examples of enriched dough include teacakes and stollen.

Pastries

- Different types of pastry are used for sweet or savoury dishes. They include:
  - shortcrust
  - choux
  - flaky/rough puff
  - suet
  - hot-water crust
  - filo pastry
  - puff pastry.

Shortcrust Pastry

- Shortcut pastry uses the rubbing-in method and is used for making pies and tarts.
- The **ratio** of ingredients is half fat to flour (1:2).
- Soft (plain) flour is used as it contains less gluten.
- Fats – a mixture of 50% lard (shortness) and 50% butter/margarine (colour and flavour) creates ideal shortening.
- Water binds the mixture. Salt adds flavour and aids the chemical reaction.
- Fat coats the flour and restricts the amount of gluten formed.
- The texture should be short, crisp and light.

Choux Pastry

- Choux pastry is used for eclairs, choux buns and gougères.
- All the ingredients must be weighed accurately.
- Strong flour is used to form the structure of the choux.
- Water and fat is brought to a **rolling boil** and the flour is added in one go immediately then mixed vigorously until the paste forms a soft ball (roux). The paste is cooled; beaten eggs are then added gradually to a **heavy dropping consistency**.
- The pastry is cooked in a hot oven.
- When the choux is baked, the water turns to steam and raises the dough and the egg protein coagulates to set the structure.

Pasta

- Pasta is made from strong wheat flour called Durum wheat, which has a high gluten content and is usually labelled type '00'.
- The liquid can be olive oil or egg. Ingredients such as spinach, tomato puree, squid ink or beetroot add colour.
- The flour and liquid are mixed then kneaded to develop the gluten content; then the dough is **rested**.
- The dough is rolled using a machine or by hand until thin and air dried. It can be made into a variety of shapes and filled.

Quick Test

1. What type of flour is used to make bread dough?
2. What gas does yeast produce?

Knife Skills

1 What **two** joints would you get from a chicken leg? Tick (✓) **two** answers. [2]

a) Breast ☐ b) Wing ☐

c) Drumstick ☐ d) Thigh ☐

2 Name and describe **two** vegetable knife cuts. [4]

Example: brunoise	Very small dice

3 Name three vegetable cuts that require you to cut the ingredient into vertical strips. [3]

Fish

1 Which of the following fish would you expect to find preserved in a can? Tick (✓) one answer. [1]

a) Cod ☐ b) Plaice ☐

c) Sardine ☐ d) Halibut ☐

2 Which choice would **not** be suitable when enrobing a fish fillet before cooking?
Tick (✓) one answer. [1]

a) Breadcrumbs ☐ b) Polenta ☐

c) Batter ☐ d) Chocolate ☐

3 Give **one** example of each of the following classifications of fish:

a) White fish .... [1]

b) Flat fish .... [1]

c) Oily fish .... [1]

4   Explain how you would select a quality fish when buying from a fresh fish counter.   [5]

_____

_____

_____

_____

_____

_____

_____

Meat

1   Which term best describes the structure of meat? Tick (✓) one answer.   [1]

   a) Meat is the flesh of animals.   ☐

   b) Meat is muscle composed of fibres held together by connective tissue.   ☐

   c) Meat is a combination of flesh and fat held together by connective tissue.   ☐

   d) Meat is a combination of tender and tough muscle fibres.   ☐

2   Which of these statements is **not** true? Tick (✓) one answer.   [1]

   a) Tender meat comes from younger animals.   ☐

   b) Tender meat has shorter muscle fibres.   ☐

   c) Tougher meat cuts can be minced to break up connective tissue.   ☐

   d) Tougher meat comes from parts of the animal that do the least work.   ☐

3   This question is about cooking meat. Name **two** methods suitable for cooking each of the following types of meat:

   a) Shin beef _____   [2]

   b) Rump steak _____   [2]

   c) Hamburger _____   [2]

4 Complete the following table for cooking meat joints. [4]

Meat	Temperature Check	Time per 500 g
Pork	75 °C or above (in England) 82 °C or above (in Scotland)	.... mins + 15 mins
Poultry	75–82 °C or above	20 mins + ... mins
Beef and Lamb	Rare: 52 °C Medium: .... °C .... : 75–80 °C	20 mins 25 mins 30 mins

5 Name **two** meats that must be cooked fully to be safe. [2]

Prepare, Combine and Shape

1 Which glaze would be most suitable for a batch of Chelsea buns? Tick (✓) one answer. [1]

a) Egg wash ☐   b) Arrowroot ☐

c) Sugar and water ☐   d) Egg yolk ☐

2 Name **two** foods from the choices below that would best demonstrate each of the food preparation methods listed in the table. [8]

apple	cheese	carrot	onion

bread	pineapple	herbs

Method	Food
Grating	
Chopping	
Slicing	
De-coring	

Dough

1 What is the most suitable flour for making **choux pastry**? Tick (✓) one answer. [1]

a) Plain flour ☐ **b)** Strong plain flour ☐

c) Self-raising flour ☐ **d)** Cornflour ☐

2 What is the ratio of fat to flour in shortcrust pastry? Tick (✓) one answer. [1]

a) 1:1 ☐ **b)** 1:2 ☐

c) 1:3 ☐ **d)** 1:4 ☐

3 What type of wheat is used to make pasta? .... [1]

4 What are the names of the proteins found in a strong flour? [2]

....

5 Name **two** foods that can be added to pasta dough to give colour. [2]

.... and ....

6 How can you make sure that pasta dough will roll out easily in a pasta machine? [2]

....

7 Complete the following instructions for making quality choux dough. Use the key word(s) to start each instruction off.

a) Weigh .... [1]

b) Add flour .... [1]

c) Add eggs .... [1]

d) Consistency .... [1]

e) Bake .... [1]

Protein and Fat

You must be able to:

- Demonstrate knowledge and understanding of the functions, structures and main sources of protein and fat
- Know the biological value of protein
- Understand the consequences of excess and deficiencies of both protein and fat.

Protein

- Protein is a macronutrient.
- Protein is formed from chains of simpler units called **amino acids**.
- Eight amino acids need to be provided by the diet and are called **essential amino acids**. Children require two additional amino acids.
- Protein is used for specific functions in the body: growth, repair, maintenance, and as a secondary energy source.

Sources of Protein

- Animal sources – meat, fish, poultry, milk, eggs, cheese, insects.
- Plant sources – soya, nuts, seeds, pulses (e.g. lentils), mycoprotein (Quorn), TVP (Texturised Vegetable Protein).

The Biological Value of Proteins

- The biological value of protein means the amount of essential amino acids present.
- Animal protein sources contain all the essential amino acids required by the body; they are of **High Biological Value (HBV)**.
- Mycoprotein (Quorn) and Texturised Vegetable Protein (TVP) are of HBV too.
- Proteins from plant sources are of **Low Biological Value (LBV)** and lack some essential amino acids. The exception is soya, which is a plant protein of HBV.

Protein Complementation

- Proteins of LBV can be eaten together to provide all the essential amino acids, for example, beans on toast. This is **protein complementation**.
- Protein complementation is important for vegetarians and vegans.

Protein Excess and Deficiencies

- Excess protein in the diet is used as energy.
- Protein deficiencies are rare but in developing countries **kwashiorkor** is a severe form of protein malnutrition.
- Some groups of people have a higher need for protein. They are:
  - babies and children – for growth
  - adolescents – for growth spurts
  - pregnant women – for the growing baby
  - nursing mothers – for lactation (milk production).

Fat

- Fat is a macronutrient. It can be solid or liquid.
- Fat is made up of **fatty acids** and **glycerol**.
- Fatty acids can be described as **saturated fats** or **unsaturated fats**.
- The structure of fatty acids influences their effect on our health.
- The characteristics of fatty acids influence cooking choice.

Functions of Fat in the Diet

- Fat provides concentrated energy.
- Fat is a source of **fat-soluble vitamins** A, D, E and K.
- Fat provides protection for the major organs in the body.
- Fat is a component of hormones.

Sources of Fat in the Diet

- Saturated fats:
  - Saturated fats can increase the **cholesterol** level in the blood.
  - Too much bad cholesterol can lead to health problems.
  - Sources of cholesterol include butter, ghee, lard, cream, hard cheese, meat pies, coconut oil and palm oil.
- Unsaturated fats:
  - Unsaturated fats can help reduce cholesterol in the blood.
  - Sources include: oily fish, nuts, seeds, avocados, vegetable oils, soya beans, and some functional foods, such as cholesterol-lowering spreads. Functional foods are foods specially developed to improve health.

Hydrogenated Fat and Trans Fats

- Making solid fat from a liquid oil is called **hydrogenation**.
- **Trans fats** can be formed when oil goes through the process of hydrogenation to form a solid. This process occurs as the molecules flip and rotate.

Fat Excess and Deficiencies

- Fat is only needed in small amounts and excess fat in the diet can lead to weight gain.
- Excess saturated fat raises blood cholesterol levels.
- Trans fats have been linked to health problems including heart disease and some cancers.
- Fat deficiency in babies and children could affect normal growth.
- Fat deficiency could result in a poor supply of fat-soluble vitamins.

 Key Point

Fats can be classified as either saturated and unsaturated. Saturated fats are considered to be more harmful to health because they raise levels of cholesterol.

Key Words

amino acids
essential amino acids
High Biological Value (HBV)
Low Biological Value (LBV)
protein complementation
kwashiorkor
fatty acids
glycerol
saturated fats
unsaturated fats
fat-soluble vitamins
cholesterol
hydrogenation
trans fats

Quick Test

1. What are the functions of fat in the diet?
2. What are the simple units of protein called?
3. Give an example of protein complementation.

Carbohydrate

You must be able to:

- Know and understand the functions, structures and main sources of carbohydrate
- Understand an individual's need for carbohydrate
- Demonstrate a knowledge and understanding of the consequences of consumption of excess carbohydrate and of deficiencies in carbohydrate.

Carbohydrate

- The body's cells require a constant supply of glucose, which is used as fuel to provide energy.
- Sugars and starches are types of carbohydrate.
- **Dietary fibre** is also a type of carbohydrate but it cannot be digested to provide energy.
- Carbohydrates are produced mainly by plants during the process of **photosynthesis**.
- Carbohydrates can be classified according to their structure: **monosaccharides**, **disaccharides**, **polysaccharides**.

> ### Key Point
>
> Carbohydrate provides the body with energy. Most of our energy should come from starchy foods.

Monosaccharides	Disaccharides	Polysaccharides
Monosaccharides are the simplest form of carbohydrate structure. They include: – Glucose – all other carbohydrate is converted into this in the body. – Galactose – found in the milk of mammals. – Fructose – found in fruit.	Disaccharides are more complex sugars that are formed when two monosaccharides join together. They include: – Sucrose – 1 unit of glucose + 1 unit of fructose. – Maltose – 2 units of glucose linked. – Lactose – 1 unit of glucose + 1 unit of galactose.	Polysaccharides are made up of many monosaccharides units joined together. They include: – Starch – many glucose units formed together. – Glycogen – formed after digestion. – Dietary fibre. – Dextrin – toasted crust on bread; sugars caramelise on the surface. – Cellulose – formed by plants from glucose. – Pectin – found in fruit, forms a gel on cooking.

Function and Sources of Carbohydrate

- Sugars are digested quickly in the body, providing instant energy.
- Starches have to be digested into sugars before absorption – this is slow energy release.
- Eating starchy foods rather than sugary foods is the healthier way to provide the body with energy. Starch (a polysaccharide) is found in bread, pasta, rice, breakfast cereals and potatoes.
- Sugars are found in a variety of sources including table sugar (sucrose), honey and jam, fruit juice, sweets and chocolate, fruit and vegetables.

Excess and Deficiencies of Carbohydrate

- Excess carbohydrate is converted to fat and is stored under the skin; this is the main cause of obesity.
- Excess sugar in the diet is linked to dental decay.
- There is evidence to suggest that the rise in Type 2 diabetes is linked to diets high in sugar.
- If insufficient carbohydrate is eaten, the body will firstly start to use protein and fat as an energy source.

Dietary Fibre

- The scientific name for fibre is Non–Starch Polysaccharide (NSP).
- Soluble NSP absorbs water, forming a gel-like substance. It can inhibit the absorption of cholesterol.
- Insoluble NSP is not absorbed by the body. It passes through the body as waste, which helps prevent bowel diseases.

Function and Sources of Dietary Fibre

- Dietary fibre makes food matter passing through the intestines soft and bulky.
- Dietary fibre can be found in wholemeal bread, wholegrain breakfast cereals (e.g. bran flakes, shredded wheat, porridge oats) wholemeal pasta and wholemeal flour; fruit and vegetables; potato skins; dried fruit, nuts and seeds, beans, peas and lentils.
- Adults should consume at least 18 g of fibre per day.
- Young children must gradually add high fibre foods to their diets.
- Fibre deficiency can lead to:
  - **Constipation** – this is when faeces become difficult to expel from the body because they are hard and small.
  - **Diverticular disease** – pouches form in the intestines, which become infected with bacteria.
- A low-fibre diet can be linked to cancer, particularly bowel cancer.

Key Words

dietary fibre
photosynthesis
monosaccharides
disaccharides
polysaccharides
Non–Starch
  Polysaccharide (NSP)
constipation
diverticular disease

Quick Test

1. What is the function of carbohydrate in the body?
2. What happens if too much carbohydrate is eaten?
3. What does NSP stand for?

Vitamins

You must be able to:

- Demonstrate knowledge and understanding of the sources and functions of both fat-soluble and water-soluble vitamins
- Know and understand the consequences of excess and deficiencies of fat-soluble and water-soluble vitamins
- Understand the retention of water-soluble vitamins during cooking.

Fat-soluble Vitamins

Vitamin	Function in the Body	Sources	Deficiency/Excess
Vitamin A (Retinol, Beta carotene)	• Normal iron metabolism • Maintenance of normal vision • Maintenance of skin and the mucus membranes • Essential for maintaining healthy immune function.	• Animal sources – liver and whole milk (retinol) • Plant sources – green leafy vegetables, carrots, and orange-coloured fruits (carotenoids) • Margarine, which is **fortified** by law	• Excess can be toxic, causing liver and bone damage • Excess retinol can lead to birth defects • Deficiency can cause night blindness
Vitamin D (Cholecalciferol)	• Absorption and use of calcium and phosphorus • Maintenance and strength of bones and teeth	• Dietary sources – oily fish, meat, eggs and fortified breakfast cereals and margarines (vitamin D added by law) • Sunlight on the skin	Deficiencies: • Bones that become weak and bend • **Rickets** in children • **Osteomalacia** in adults • Weak teeth
Vitamin E (Tocopherol)	• **Antioxidant** that helps protect cell membranes • Maintains healthy skin and eyes	• Polyunsaturated fats, e.g.sunflower oils • Nuts, seeds and wheatgerm	• Very rare
Vitamin K (Phytomenadione)	• Normal clotting of blood	• Green leafy vegetables, cheese, bacon and liver	• A deficiency is rare • Newborn babies are given a dose of vitamin K

Water-soluble Vitamins

Vitamin	Function in the body	Sources	Deficiency
Vitamin B1 (thiamin)	• Release of energy from carbohydrates • Nervous system function • Normal growth of children	• Wholegrain products • Meat • Milk and dairy • Nuts • Marmite • Fortified breakfast cereals • Fortified white and brown flour	• Beri-beri, which affects the nervous system

Vitamin B2 (riboflavin)	• Energy release from foods • Healthy nervous system	• Same as vitamin B1 thiamin	• Cracking skin around the mouth • Swollen tongue • Failure to grow
Folate (Folic acid)	• Neural tube development in unborn babies	• Green leafy vegetables • Potatoes, asparagus, bananas • Beans, seeds and nuts • Wholegrain products • Breakfast cereals	• **Spina bifida** in unborn babies. Pre conception and pregnant women need a good supply
Vitamin B12 (cobalamin)	• Supports production of energy • Protective coating around nerve cells	• Meat and fish • Cheese, eggs and milk • Marmite • Fortified breakfast cereals	• Pernicious anaemia
Vitamin C – (ascorbic acid)	• Absorption of iron • Production of collagen that binds connective tissue • Antioxidant – protects from pollutants in the environment	• Citrus fruits • Kiwi fruit • Blackcurrants • Potatoes • Salad and green vegetables, e.g. broccoli, kale	• Weak connective tissue and blood vessels • Bleeding gums and loose teeth • Anaemia • Severe cases – scurvy

Vitamins and Food Preparation

- Vitamins B and C are water-soluble so will dissolve into water during cooking. Vitamins B1 and C are destroyed by heat. Vitamin C is destroyed on exposure to oxygen.
- To maximise vitamin retention:
  - prepare foods quickly just before serving
  - use small amounts of boiling water to cook
  - use cooking liquid to make sauces
  - avoid lots of cutting of vegetables.
  - water soluble vitamins – avoid cooking in water, instead steam, roast, fry or grill.
  - fat-soluble vitamins – avoid cooking in fat, instead boil, steam or grill.
- Vitamins A, C and E contain antioxidants which work together to protect cells against oxidative damage from free radicals.

Key Point

Vitamins are micronutrients, required in small amounts to do essential jobs in the body.

Key Point

Water-soluble vitamins are easily destroyed during preparation and cooking.

Key Words

fortified
rickets
osteomalacia
antioxidant
thiamin
riboflavin
spina bifida
cobalamin
ascorbic acid

Quick Test

1. What are the fat-soluble vitamins?
2. Give **three** good sources of vitamin C.
3. Which vitamin is connected with the healthy development of the spine in unborn babies?

Minerals and Water

Food Nutrition and Health

You must be able to:

- Demonstrate knowledge and understanding of the functions and main sources of calcium, iron, sodium, iodine and fluoride
- Show knowledge and understanding of the consequences of deficiencies of calcium, iron, sodium, iodine and fluoride in the diet
- Know and understand the function of water in the diet.

Minerals

- Minerals are micronutrients.
- Minerals include calcium, iron, fluoride, sodium and iodine.

> ### Key Point
>
> Minerals are required in small amounts and have a variety of essential functions in the body.

Calcium

Function	Sources	Deficiency
• Strengthens bones and teeth – with Vitamin D • Bones are able to reach **peak bone mass** – maximum strength • Growth of children • Clots blood after injury • Promotes nerves and muscles to work properly.	• Milk and dairy foods • Green leafy vegetables • White bread – calcium is added by law • Soya products. • Fish eaten with the bones, e.g. sardines 	• Bones don't reach peak bone mass and become weak and break easily – common in older people (osteoporosis). • During pregnancy, a woman's teeth and bones weaken. • Poor clotting of the blood.

Iron

Function	Sources	Deficiency
• Supports the production of **haemoglobin** in red blood cells; this transports oxygen around the body • Vitamin C is required to absorb iron.	• Red meat – liver • Lentils, dried apricots, cocoa, chocolate, corned beef and curry spices • Green leafy vegetables, e.g. spinach • Breakfast cereals fortified with iron.	• Iron deficiency is called **anaemia**; symptoms include tiredness • Low intake of dietary iron, particularly in young women, can cause iron deficiency anaemia • Pregnant women may become deficient due to additional blood needed to support the growing baby.

Fluoride

Function	Sources	Deficiency
• To strengthen the enamel layer of teeth.	• Saltwater fish, tea • Some water authorities fortify their water supply with fluoride.	• Teeth may develop cavities and require filling.

Sodium

Function	Sources	Excess and Deficiency
• Needed by the body to regulate the amount of water in the body • Needed to assist the body in the use of energy • Required to help control muscles and nerves.	• Processed foods – for flavour and as a preservative • Salt added to food in home cooking for flavour • Salt added at the table.	• A deficiency is rare, but can lead to muscle cramps after exercise in hot conditions • People suffering with sickness and diarrhoea can lose salt • Excess salt in the diet is linked to high blood pressure, heart disease and strokes.

Iodine

Function	Sources	Deficiency
• Regulates hormones in the thyroid, which controls the body's metabolic rate.	• Seafood • Foods grown in iodine-rich soils.	• Poor functioning of the thyroid gland, resulting in feeling tired and lethargic • A goitre (swelling) in the thyroid gland.

Water

- The body is nearly two-thirds water. The functions of water are:
  - For normal brain function
  - To decrease risk of kidney problems
  - For normal blood pressure
  - To help bowel movements
  - To regulate temperature and maintain hydration
  - To make body fluids – blood, saliva, mucus membranes.
- The main sources of water are drinking water, milk, tea, coffee and fruit juices.
- The remaining 20% of water needed comes from foods such as soup, yoghurt, fruit and vegetables.
- Water should be drunk every day. 6–8 glasses daily is recommended.
- People with increased water needs are:
  - Those with an active lifestyle during hot weather – water needs increase due to perspiration.
  - Anyone suffering vomiting or diarrhoea, to avoid **dehydration**.
  - **Lactating** mothers (breast feeding), for milk production.
  - Elderly people, to help prevent kidney problems and infections.
  - During hot weather water needs increase due to perspiration.

Key Point

Water makes up two thirds of the body so it is vital to drink regularly to stay hydrated.

Key Words

peak bone mass
haemoglobin
anaemia
thyroid
dehydration
lactating

Quick Test

1. Why do teenage girls need an increased supply of iron?
2. What is peak bone mass?
3. Which hormones in the body are affected by a lack of iodine?
4. Name **two** good sources of calcium in the diet.

Making Informed Choices

You must be able to:

- Demonstrate knowledge and understanding of the Eatwell Guide and the Healthy Eating Guidelines
- Have knowledge and understanding of the diet requirements throughout life.

The Eatwell Guide

- The **Eatwell Guide** shows the proportions of food groups that should be eaten daily in a well-balanced diet.

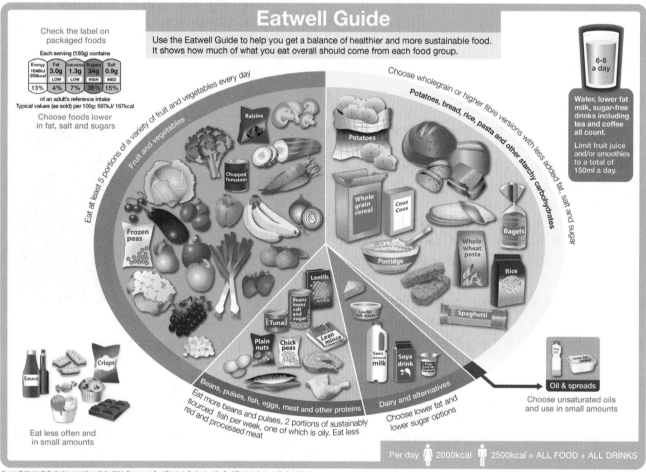

Eatwell Guide

Use the Eatwell Guide to help you get a balance of healthier and more sustainable food. It shows how much of what you eat overall should come from each food group.

Check the label on packaged foods

Each serving (150g) contains

Energy 1046kJ 250kcal	Fat 3.0g	Saturates 1.3g	Sugars 34g	Salt 0.9g
	LOW	LOW	HIGH	MED
13%	4%	7%	38%	15%

of an adult's reference intake
Typical values (as sold) per 100g: 697kJ / 167kcal

Choose foods lower in fat, salt and sugars

Eat at least 5 portions of a variety of fruit and vegetables every day

Choose wholegrain or higher fibre versions with less added fat, salt and sugar

Potatoes, bread, rice, pasta and other starchy carbohydrates

6-8 a day

Water, lower fat milk, sugar-free drinks including tea and coffee all count.

Limit fruit juice and/or smoothies to a total of 150ml a day.

Beans, pulses, fish, eggs, meat and other proteins
Eat more beans and pulses, 2 portions of sustainably sourced fish per week, one of which is oily. Eat less red and processed meat

Dairy and alternatives
Choose lower fat and lower sugar options

Oil & spreads
Choose unsaturated oils and use in small amounts

Eat less often and in small amounts

Per day 👩 2000kcal 👨 2500kcal = ALL FOOD + ALL DRINKS

Source: Public Health England in association with the Welsh Government, Food Standards Scotland and the Food Standards Agency in Northern Ireland

© Crown copyright 2016

Current Guidelines for a Healthy Diet

- Base your meals on starchy carbohydrates.
- Eat lots of fruit and vegetables (5–7 portions per day).
- Eat plenty of fish, including oily fish.
- Cut down on saturated fat and sugars.
- Eat less salt – no more than 6g a day for adults.
- Get active and be a healthy weight.
- Don't get thirsty (drink six to eight glasses of water a day).
- Don't skip breakfast.

Key Point

Nutritional needs change throughout life, but everyone needs to consider the current healthy eating guidelines when planning meals.

Nutritional Needs Throughout Life

Babies:

- Newborn babies should have only milk for the first 4–6 months of life.
- First milk is called **colostrum** and it is full of antibodies.
- Human milk provides babies with all their nutritional requirements, except for iron. Babies are born with a supply of iron stored in the liver.
- The introduction of solid foods is called 'weaning'.

Children

- Toddlers aged 1–3 years grow rapidly and so it is essential they get a well-balanced diet to support their physical development.
- Toddlers are physically very active and need a good supply of fat, which also helps the development of the brain and nervous system.
- New healthy foods need to be introduced in an attractive and appealing way.
- Sweets, biscuits, cakes and fizzy drinks must be avoided. Sugary foods cause dental decay and are strongly linked to obesity.

Teenagers

- Adolescence is a period of rapid growth and this is when **puberty** occurs.
- The need for energy and most nutrients is relatively high.
- After **menstruation** begins, girls need more iron to replace blood losses. **Iron-deficiency anaemia** is common in teenage girls.

Pregnancy

- A healthy diet is required in pregnancy to ensure the baby receives the essential nutrients required for development.
- Folate (Folic acid) is needed both before and during early pregnancy for the development of the neural tube of the **foetus**.
- The baby's bones require a good supply of calcium from the mother's diet.
- A pregnant mother will need a diet rich in iron for the production of additional blood supply and to lay down an iron store in the baby's liver.
- Constipation is common in pregnancy so a diet high in dietary fibre (NSP) is important.

Older people

- In older people energy requirements decrease. They will require smaller portions at meal times.
- Older people need to keep hydrated by drinking plenty of fluids.
- **Osteoporosis** may occur when bones become weak, brittle and break easily.
- Older people are advised to eat calcium-rich foods to help strengthen bones.

Quick Test

1. Give **three** recommended guidelines to a healthy diet.
2. Why is a good supply of folate (folic acid) needed in early pregnancy?
3. What is colostrum?
4. In babies, what is the introduction of solid foods called?
5. What is osteoporosis?

Key Words

Eatwell Guide
colostrum
puberty
menstruation
iron-deficiency anaemia
foetus
osteoporosis

Diet, Nutrition and Health

You must be able to:

- Demonstrate knowledge and understanding of the health conditions linked to food and nutrition
- Demonstrate knowledge and understanding of the terms relating to diet, health and energy needs.

Diet-related Medical Conditions

Bowel Cancer

- Bowel cancer is the second-biggest illness/cause of death in the UK.
- The risk of bowel cancer and diverticular disease can be greatly reduced by increasing fibre/NSP (non-starch polysaccharide) intake.

Obesity, Coronary Heart Disease and Type 2 Diabetes

- **Obesity** is an abnormal accumulation of body fat. Obesity is associated with coronary heart disease, diabetes, types of cancer and high blood pressure.
- **Coronary heart disease** is caused by fatty substances (cholesterol) building up in the walls of the arteries that run to the heart.
- Arteries narrow, reducing the supply of oxygen to the heart.
- Diets high in saturated fats produce cholesterol. Everyone should reduce their intake of saturated fat.
- High blood pressure can lead to an increased risk of stroke. High blood pressure can be linked to increased salt intake. The recommended amount of salt to be consumed per day is 6 grams. Salt can be hidden in processed foods. We are recommended to reduce our salt intake.
- With **Type 2 diabetes** either too little insulin is produced, or the body's cells fail to react to the insulin that is produced. This results in high levels of sugar in the blood.
- Diabetes is controlled by careful management of sugar in the diet, plus insulin medication frequently injected.
- Type 2 diabetes is increasingly linked to obesity.

Iron Deficiency Anaemia

- Anaemia is common in teenage girls due to menstruation.
- Symptoms of anaemia include tiredness, lack of energy, shortness of breath and pale complexion.
- Pregnant women also need increased iron supplies in the diet.
- Anaemia is simply diagnosed with a blood test.

Dental Health and Bone Health

- Sugar is a major cause of **tooth decay** in children.
- Sugar increases acids on the teeth, causing irreparable damage.
- In tooth decay, acids erode the protective white enamel surface of the teeth.

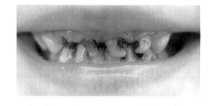

> **Key Point**
>
> Obesity is the main cause of diet-related illness in the UK.

> **Key Point**
>
> Osteoporosis means 'porous bones' – the bones lose their strength and are more likely to break.

- **Osteoporosis** is common in old age. Bones can become weak, brittle and more likely to break. It also causes the spine to curve forward, making walking difficult.
- A diet rich in calcium and vitamin D is required to ensure maximum bone strength.

Energy Needs

- Energy is required for us to grow, to keep the basic functions of our body going, and to be physically active.
- Energy requirements depend on your **Basal Metabolic Rate (BMR)**. BMR is the energy needed by the body to power your internal organs when completely at rest.
- An individual's BMR depends on their age, gender, and body size.
- We use energy for movement of all types, known as **Physical Activity Level (PAL)**. More physically active people require more food to supply their energy needs.
- Nutritionists devise **Estimated Average Requirements (EARs)** tables that provide guidelines to energy needs at various stages of life.
- Malnutrition is a result of under consumption of nutrients. Anorexia and bulimia can lead to malnutrition symptoms.

Energy Intake and Expenditure

- The amount of energy **calories (kcal) or kilojoules (kJ)** a food contains per gram is known as its **energy density**.
  – Fat = 9 kcal/g; Protein = 4 kcal/g; Carbohydrate = 4 kcal/g
- What we weigh depends on the balance between how much energy we consume from our food and how much energy we use up by being physically active.
- Government guidelines state that we need to undertake 60 minutes of aerobic activity every day.
- **Reference Intake (RI)** is the recommended amount of each nutrient that is required to eat daily. These replaced GDA in some instances.

Body Mass Index

- **Body Mass Index (BMI)** is a measure that adults can use to see if they are a healthy weight for their height.
- The ideal healthy BMI is between 18.5 and 25.

Key Point

Energy balance is the balance of energy consumed through eating and drinking compared to energy burned through physical activity.

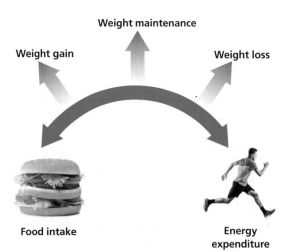

Weight maintenance

Weight gain

Weight loss

Food intake

Energy expenditure

Key Words

Basal Metabolic Rate (BMR)
Physical Activity Level (PAL)
Estimated Average Requirements (EARs)
energy density
Reference Intake (RI)
Body Mass Index (BMI)

Quick Test

1. What is osteoporosis?
2. Name **two** symptoms of anaemia.

Review Questions

Knife Skills

1 A chef needs to have their own set of knives. Name and describe **three** different chef knives, giving one possible use for each. [9]

Name of knife	Description	One Possible Use

Fish

1 What sign would indicate that a fish is **not** good quality? Tick (✓) **one** answer. [1]

a) Bright eyes ☐ b) Thin layer of clear slime ☐

c) Firm flesh ☐ d) Offensive odour ☐

2 Which of the following statements is **true**? Tick (✓) **one** answer. [1]

a) Fish takes a long time to cook. ☐

b) Fresh fish should have red gills. ☐

c) Fish cannot be fried without using a coating. ☐

d) Fish contains a lot of carbohydrate. ☐

3 Explain **three** ways in which fish can be preserved commercially and give **one** example of each. [6]

Method	Example

4 Which **two** methods of preserving fish involve removing moisture from the fish? [2]

5 Name **four** nutrients found in oily fish. [4]

1 .... 3 ....

2 .... 4 ....

Meat

1 What temperature should pork be cooked to in order for it to be safe to eat? Tick (✓) **one** answer. [1]

a) 63 °C ☐ b) 72 °C ☐ c) 74 °C ☐ d) 75 °C ☐

2 What causes meat to become firm when cooked? Tick (✓) **one** answer. [1]

a) Non-enzymic browning ☐ b) Coagulation ☐

c) Maillard reaction ☐ d) Gelatinisation ☐

3 Explain which type of meat comes from these animals:

a) Pig .... [1]

b) Cattle .... [1]

c) Sheep .... [1]

d) Deer .... [1]

e) Poultry .... [1]

4 Name **three** nutrients that are found in meat. [3]

_____

_____

5 Explain **three** ways to ensure that raw meat is stored safely. [6]

_____

_____

_____

Prepare, Combine and Shape

1 What type of binding agent is generally used in a sausage mixture? Tick (✓) **one** answer. [1]

a) Breadcrumbs ☐   **b)** Flour ☐

c) Egg ☐   **d)** Water ☐

2 What is the correct term given for the vegetables shown? Tick (✓) **one** answer. [1]

a) Jardinière ☐   **b)** Julienne ☐

c) Macedoine ☐   **d)** Brunoise ☐

3 Which of these dishes uses **three** methods of combining, shaping and coating? Tick (✓) **one** answer. [1]

a) Burgers ☐   **b)** Fishcakes ☐

c) Chicken goujons ☐   **d)** Koftas ☐

4 What kind of glaze is used on a batch of scones? [1]

_____

5 Give **two** examples of foods that can be shaped using a mould. [2]

_____ and _____

6 What is the advantage of using cutters and a rolling pin when making a batch of biscuits? [2]

_____

Dough

1 Which of the following dishes would you **not** make using choux pastry? Tick (✓) **one** answer. [1]

a) Eclairs ☐   b) Gougère ☐

c) Choux buns ☐   d) Bakewell tart ☐

2 Complete the following questions to show your understanding of shortcrust pastry.

a) What is the ratio of fat:flour? _____ [1]

b) What type of flour should be used? _____ [1]

c) What fats should be used?

_____ [2]

d) What method should be used? _____ [1]

e) What texture should the cooked pastry have? _____ [1]

3 What are the functions of each of the following ingredients in bread dough?

a) Yeast _____ [1]

b) Sugar _____ [1]

c) Fat _____ [1]

d) Warm liquid _____ [1]

4 Why does pasta dough need to be 'rested' after kneading? [2]

_____

_____

5 Which type of wheat flour is best for making a pasta dough? Tick (✓) **one** answer. [1]

a) White ☐   b) Wholemeal ☐

c) Durum ☐   d) Granary ☐

Practice Questions

Protein and Fat

1 What are the main functions of protein in the diet? Tick (✓) **one** answer. [1]

    a) Healthy teeth and strengthening bones ☐

    b) To support the absorption of iron ☐

    c) Growth, maintenance and repair ☐

    d) For a healthy digestive system ☐

2 Which of the following is a good source of High Biological Value protein? Tick (✓) **one** answer. [1]

    a) Lentils ☐   **b)** Cheese ☐

    c) Baked beans ☐   **d)** Bread ☐

3 Give **one** example of how protein foods can be combined to complement each other. [2]

_____

4 Which of the following is a function of fat in the diet? Tick (✓) **one** answer. [1]

    a) Repair of body cells ☐   **b)** Insulation and warmth ☐

    c) Protects enamel on teeth ☐   **d)** Makes connective tissue ☐

5 Name **two** foods which contain saturated fatty acids. [2]

_____

Carbohydrate

1 What is the name given to the production of carbohydrates by plants? Tick (✓) **one** answer. [1]

    a) Cholesterol ☐   **b)** Hydrogenation ☐

    c) Photosynthesis ☐   **d)** Marinade ☐

2 Give **three** good sources of starchy carbohydrates. [3]

_____

_____

3 Explain why the body needs carbohydrates. [2]

_____

_____

_____

4 Dietary fibre is also known as: [1]

N _____ S _____ P _____ .

5 Give **three** good sources of dietary fibre in the diet. [3]

_____

_____

Vitamins

1 Which are the water-soluble vitamins? [2]

_____

2 Which food is a good source of vitamin C? Tick (✓) **one** answer. [1]

a) Bread ☐   **b)** Oily fish ☐   **c)** Oranges ☐   **d)** Butter ☐

3 Explain how the loss of water-soluble vitamins can be reduced when preparing and cooking fresh vegetables. [4]

_____

_____

_____

_____

4 What type of vitamins are vitamins, A, D, E and K? [1]

_____

Practice Questions

5 Apart from dietary sources, how else does the body obtain vitamin D? [2]

Minerals and Water

1 Name the condition caused by a lack of iron in the diet. Tick (✓) **one** answer. [1]

a) Scurvy ☐   **b)** Anaemia ☐   **c)** Beri-beri ☐   **d)** Dermatitis ☐

2 Milk contains calcium. Name **two** other good sources of calcium in the diet. [2]

3 What is the function of iron in the body? [2]

4 Name **two** good sources of iron in the diet. [2]

5 Give **three** reasons why water is an important part of a healthy diet. [3]

Making Informed Choices

1 Complete the sentence. [1]

The E .... G .... shows the types and proportions of the main food groups that are needed for a healthy balanced diet.

2 Eating less salt and less fat are two of the current healthy eating guidelines. List **three** other healthy eating guidelines. [3]

....

....

3 What term is used to describe the gradual introduction of solid foods into a baby's diet? [1]

....

4 What does the yellow group on the Eatwell Guide show? [1]

....

Diet, Nutrition and Health

1 Which group of people require an increased supply of iron? Tick (✓) **one** answer. [1]

    a) New-born babies ☐   **b)** Teenage girls ☐

    c) Elderly men ☐   **d)** Someone working in an office. ☐

2 What is the maximum daily recommended amount of salt? Tick (✓) **one** answer. [1]

    a) 2 g ☐   **b)** 6 g ☐   **c)** 16 g ☐   **d)** 24 g ☐

3 Obesity is linked to many other health conditions. Name **two** of these conditions. [2]

....

4 People with Type 2 diabetes cannot produce enough of which hormone? [1]

....

5 What are the health risks of a high-fat diet? [3]

....

....

6 Explain what is meant by basal metabolic rate (BMR). [2]

....

....

Cooking of Food, Heat Transfer and Selecting Appropriate Cooking Methods

You must be able to:

- Know and understand the reasons why food is cooked and how heat is transferred to food
- Know the reasons for selecting different cooking methods.

Why is Food Cooked?

- Food is cooked in order to:
  - make it safe to eat (cooking destroys micro-organisms)
  - change raw food to cooked food
  - make it palatable: develop flavours; improve mouthfeel; improve texture; reduce bulk; improve colour
  - help keep quality (extend shelf life)
  - make it easier to digest
  - give variety to diet.

How is Food Cooked?

- Cooking uses heat to change texture, flavour and colour of food.
- Cooking methods are **wet** (moist), **dry**, or **fat-based**:
  - Wet methods are boiling, steaming, stewing, poaching, casseroling and braising.
  - Dry methods are baking, roasting and grilling.
  - Fat-based methods are frying and stir-frying.
- Pre-cooking methods can improve tenderness and flavour, for example, a marinade for meat or fish.
- The selection of a cooking method depends on:
  - the type of food being cooked
  - the time available
  - the skill of the cook and the facilities available
  - the need to achieve desired characteristics, e.g. browning
  - the need to conserve vitamins, e.g. steam rather than boil to prevent loss of vitamins
  - the desire to improve palatability, which affects the appearance, colour, flavour, texture and smell.

 Key Point

Cooking food makes it safe, allows it to keep for longer and makes it more palatable.

Microwave Cooking

- Microwave cooking uses a type of radiation called microwaves, which travel in straight lines and penetrate the food.
- The microwaves vibrate water molecules creating friction, which makes heat.
- **Hot spots** can occur, so food should be stirred and left to stand to allow the heat to be distributed evenly.
- Microwaves alone do not brown or crisp foods.

Summary of Cooking Methods

Method	Characteristics	Examples
Water-based		
Boiling	softens	vegetables, rice
Simmering	tenderises, evaporates	stews, sauces, curry
Poaching	tenderises	fish, chicken, eggs
Braising	tenderises, softens	meats, fish, vegetables
Steaming	tenderises	vegetables
Oven		
Moist or fat	tenderises	vegetables, joints of meat, potatoes
Dry		
Grilling	chars, browns, crisps	bacon, toast, cheese
Baking	browns, crisps	cakes, pastries
Dry frying	browns, crisps	bacon, lardons, chorizo, nuts
Roasting	browns, crisps	roast chicken, potatoes
Fat-based		
Deep frying	browns, crisps	battered fish
Shallow frying	sets, browns	eggs, onions
Stir fry	softens, reduces bulk	bean sprouts

Heat Transfer

- There are three ways in which heat is transferred through food:

Conduction	Convection	Radiation
• Food molecules vibrate to transfer heat via **conduction**. • Heat is transferred by contact of heat source to pan to food, e.g. frying.	• **Convection** takes place in air (in ovens) or liquids. • Currents occur as heated air or water rises and cooler air or water falls, e.g. boiling water.	• Heat energy passing in direct lines to the food, e.g. grill to food. • Energy from microwaves penetrate food to transfer energy.

Key Point

Cooking methods can achieve specific characteristics in food.

Key Point

Heat is transferred by conduction, convection and radiation. Cooking commonly uses a combination of heat transfer methods.

Conduction

Convection

Radiation

Key Words

palatability
microwave
radiation
conduction
convection

Quick Test

1. Cooking food makes it more palatable. Is this true or false?
2. What is the main heat transfer method when boiling food?
3. How can cooking methods be classified?
4. Name **three** types of heat transfer.

Proteins and Enzymic Browning

You must be able to:

- Understand protein denaturation and protein coagulation
- Know about the properties of protein in gluten formation
- Understand enzymic browning and oxidation in fruit and vegetables.

Protein Denaturation

- **Denaturation** occurs when the structure of amino acids found in protein are altered. They change shape or unfold because chemical bonds are broken.
- Protein in foods can be denatured (altered) by heat, reduction of **pH level** (more acid), enzymes and mechanical actions. See chart below.

> ### Key Point
>
> Proteins are denatured during cooking. Egg proteins coagulate or set when they are heated.

Heat	• Cooking denatures proteins. • Denaturation occurs when the structure of amino acids found in protein change shape after cooking – the protein molecule uncoils when cooked.
pH	• pH is the level of acidity or alkalinity in a food. • pH is measured from 1, which is very acid, to 14, which is very alkaline. • Reducing the pH by using lemon juice or vinegar in a **marinade** denatures the protein in foods to make them tender, tasty and moist.
Enzymes	• Meat tenderisers cause protein denaturation, making meat more tender. • Enzymic tenderisers are in the form of papain and bromelain. Papain can be found naturally in papaya and bromelain can be found in fresh pineapple. • Acidic pH from yogurt, buttermilk, vinegar or citrus fruits helps to tenderise fish or meat (marinades).
Mechanical Actions	• Whisking egg white denatures the protein by uncoiling and unfolding, e.g. foam formation occurs (gas in liquid).

Protein Coagulation

- Protein coagulation is a type of protein denaturation.
- Examples of coagulated foods are egg custards and quiches.
- It causes a change in texture, for example, runny eggs become coagulated (set).
- It usually starts at 60°C and is completed by 70°C.
- It is irreversible and causes loss of solubility.

Gluten Formation

- Gluten formation occurs when water is added to a wheat flour to form a dough. Wheat flour contains two proteins, glutenin and gliadin, which combine to form gluten.
- Strong wheat flour for bread making contains more gluten than plain flour.

Gluten in Bread

- Gluten makes dough stretchy and elastic.
- Salt and kneading help strengthen gluten.
- Gluten forms the structure of a baked loaf of bread.

Gluten in Pastry

- Rubbing fat into flour makes short gluten strands – the scientific term for this is shortening.
- Gluten forms the structure in baked pastry.

Gluten in Pasta Making

- Gluten in wheat flour helps pasta hold its shape, e.g. fusilli.
- Gluten also makes the pasta dough flexible and increases its ability to hold various shapes.

Proteins and Enzymic Browning

- Enzymic browing occurs on the surface of cut fruits, such as apples, and on the surface of cut vegetables, such as potatoes.
- It happens due to cell enzymes reacting with air (oxidation).
- Enzymic browning can be prevented (inhibited) by:
  - blanching cut fruits or vegetables in boiling water
  - blanching vegetables before freezing, which inhibits the enzymic action that can discolour food
  - dipping fruit or vegetables in acid, e.g. lemon juice
  - removal of air by submerging in water
  - cooking, e.g. stewing fruit, roasting parsnips.

Oxidation

- Oxidation causes discolouration, e.g, cut lettuce leaves turn pink-brown.
- Oxidation also causes vitamins to be lost, particularly vitamin C.
- Oxidation enables enzyme activity, e.g. browning, discolouration.
- Oxidation can be reduced during preparation and cooking by:
  - cooking vegetables in small amounts of water
  - using a quicker, shorter method of cooking, e.g. steaming, stir fry
  - serving vegetables immediately after cooking
  - keeping the lid on when boiling vegetables
  - using the cooking water in gravy.

> **Key Point**
>
> Wheat flour contains the protein gluten. Gluten forms the structure of pastries, breads and cakes.

> **Key Point**
>
> Enzymes can cause the browning of fruit and vegetables. Fruit and vegetables need careful handling during preparation to prevent enzymic browning.

> **Key Words**
>
> denaturation
> pH level
> marinade
> enzymic browning
> oxidation

> ### Quick Test
>
> 1. What causes the browning of cut fruit or vegetables?
> 2. What is the term used to explain the way heat changes the texture of egg proteins?

Carbohydrates

You must be able to:

- Understand the functional and chemical properties of carbohydrates, which are gelatinisation, dextrinisation and caramelisation.

Gelatinisation

- **Gelatinisation** occurs when starches (wheatflour, cornflour or arrowroot) thicken liquids. The process needs heat and agitation (stirring), especially in sauce making. It occurs during the cooking of starchy food such as potatoes, rice or pasta.

Gelatinisation in Sauce Making

- When making a sauce gelatinisation occurs.
- Starch grains absorb liquid, swell, burst (at 80°C) and finally thicken a sauce. Starches complete thickening at just under 100°C, so it is important to cook to boiling point to avoid a sauce with a raw taste.
- Sauces need stirring (agitating) to prevent lumps forming, to stop the sauce from sticking to the bottom of the pan and burning, and to help the process of gelatinisation.
- Basic recipes use different methods and different thickening agents, as seen when using the roux method, the blending method and the all-in-one method.

Gelatinisation and Ratios

- A sauce is viscous, which means it can either be poured (e.g. parsley sauce) or used to coat an ingredient (e.g. cauliflower cheese) or to bind other dry ingredients together (e.g. rissoles).
- The change in **viscosity** is due to the ratio of thickening agent to liquid; more starch gives a thicker sauce. The ratio selected changes the **consistency** of the sauce from runny to thick.
- Retrogradation is the deterioration of a starch-based sauce on keeping – this results in shrinkage, drying and cracking.
- Synerisis is the loss of fluid from a foam or set mixture, e.g. lemon meringue pie, and in cheese making.

Modified Starches and Gelatinisation

- Modified starches are used to help gelatinisation occur in different ways. Quick-cook pasta or rice are modified by pre-gelatinisation.
- Milkshakes use starch modified to allow cold liquid thickening.
- Instant thickening granules are modified starch that can be sprinkled into boiling liquids.

> **Key Point**
>
> Gelatinisation is the function of starches as thickening agents.

- Modified starches thicken cold desserts without the application of heat, and are used to thicken and stabilise salad dressings.

Dextrinisation

- **Dextrinisation** occurs when starch is toasted or cooked by dry heat, e.g. toasted or charred bread.
- It is a result of starch breakdown by dry heat to form dextrins.
- It changes the properties of starch as a result of heat application.
- Dextrinisation is known as non-enzymic browning.
- Dextrins taste sweeter than starch and add flavour to toasted, charred or baked goods.
- Dextrins are hygroscopic, absorbing moisture from the air, e.g. toasted or baked products soften slightly on keeping. Baked produce is best stored in an airtight tin or container with lid.
- Characteristics of dextrinisation are golden colours, browning, sweeter taste and crispness.

Caramelisation

- **Caramelisation** causes sugar to change colour and flavour due to dry or moist heat.
- It causes surface browning on baked goods containing sugar.
- It changes the properties of sugar; solutions become syrups.
- It is known as non-enzymic browning.
- Characteristics of caramelisation are a golden colour, browning, gloss, sweetness and stickiness.

Examples of Caramelisation (Browning) in Food Preparation

- Frying onions, or frying or roasting parsnips, potatoes or squash.
- Making a crème brulee or crème caramel.
- Spun sugar caramelisation is created through the application of **dry** heat to sugar.
- Fudge, toffee or halva are created through the application of **wet** heat to caramelise the sugar.
- Preparing glazed pork ribs or chicken wings.

Browning and Carbohydrates

- Dry methods of cooking gradually turn starches and sugars golden brown. This aids the palatability of the products by improving flavour, texture and colour.
- Browning is often used as an indicator of adequate cooking e.g. cook until golden brown.

Key Point

Sauces can be different thicknesses when the proportion of ingredients is altered.

Key Point

Dextrinisation is the term used to describe browning of starch caused by heat. Caramelisation is the browning of sugars caused by heat.

Key Words

gelatinisation
viscosity
consistency
dextrinisation
caramelisation

> **Quick Test**
>
> 1. What term describes thickening a sauce using starch?
> 2. What sort of heat transfer commonly causes dextrinisation?
> 3. Dextrins attract and absorb moisture from the air. Is this true or false?

Fats and Oils

You must be able to:

- Understand the role of fats and oils in shortening, plasticity, aeration and emulsification
- Know the functional and chemical properties of fats and oils.

Types of Fat

- Fat from animal sources include butter and lard – these are saturated fats.
- Fat from vegetable sources include margarine and vegetable shortening – these are unsaturated fats and are suitable for vegetarians, vegans and certain religions.

Shortening

- **Shortening** is a process using fat that creates a characteristic short, crumbly texture.
- Shortening is seen in shortcrust pastry, shortbread biscuits, cookies and rich pastries.
- The process in food preparation most likely to bring about shortening is the rubbing-in method.

How Does a Fat Shorten a Pastry Mixture?

- Fats with plasticity are good shortening agents because they rub-in easily.
- Fat coats the flour grains, preventing gluten development.
- The cooked texture is short and crumbly.

What Happens During the Cooking of Pastry?

- The flour grains absorb the fat.
- Pastry changes from pliable to rigid (the gluten sets).
- Pastry turns golden brown.

Plasticity

- **Plasticity** means the ability of a fat to change properties over a range of temperatures. Temperature is an important factor in plasticity of fats.
- Cold fats are solid and firm.
- Fats at room temperature become spreadable and soft.
- Plastic fats such as butter or margarine can be used for spreading, rubbing-in, creaming, melting-method cooking or for muffins.
- Warmed fats melt and become runny.

> **Key Point**
>
> Fats make pastry short and crumbly. Fats give colour and flavour to pastry. The plasticity of fat allows it to be used for rubbing-in, spreading and creaming.

Aeration

- **Aeration** helps products have a light and open texture.
- Fats aerate mixtures during beating or **creaming** with sugar.
- Aeration increases the volume of a product by incorporating air.
- Beating, whipping, creaming and whisking are methods that help aeration.
- During preparation of a creamed mixture:
  - the fat and sugar are creamed together, trapping air (aeration)
  - the mixture becomes paler
  - an air-in-fat **foam** is formed.
- During baking:
  - trapped air expands
  - the cake rises.

Emulsions

- Emulsions are mixtures of liquids that do not normally mix (known as 'immiscible' liquids), e.g. oil and water.
- Emulsifiers have a hydrophilic end, which is water-loving and forms chemical bonds with water, and a hydrophobic end, which is water-hating and forms chemical bonds with oil.
- Fats and oils add texture, flavour and colour to emulsified sauces such as Hollandaise, which is a hot emulsion, and mayonnaise, which is a cold emulsion.
- Stabilisers keep emulsions mixed, preventing them separating.

types of emulsions

oil-in-water   water-in-oil

The Process of Emulsification

- **Emulsification** requires agitation by whisking, by mixer or food processor.
- It requires slow addition of oil to prevent the emulsion splitting.
- Emulsification is helped by a natural emulsifier called lecithin, which is present in egg yolks.
- A good example of emulsification is seen when making mayonnaise, a smooth, stable emulsion for salad dressing.
- Vinaigrette dressing (oil and vinegar) is not a stable emulsion – it will separate out on standing.

Key Point

Fats can help aeration in baking.

The creaming method aerates a cake mixture and helps it to rise.

Oil and balsamic vinegar

Key Point

Emulsions are mixtures of liquids that do not normally mix, e.g. oil and water. Egg yolks contain lecithin, a natural emulsifier. Eggs help stabilise mayonnaise.

Key Words

shortening
plasticity
aeration
creaming
foam
emulsification

Quick Test

1. What term describes how fat makes a short texture product?
2. The ability of a fat to change properties is known as plasticity. Is this true or false?
3. Which basic cake-making process traps air into the cake?

Raising Agents

You must be able to:

- Understand the processes of raising or aerating using physical and mechanical methods
- Know and understand the working properties of chemical and biological raising agents.

Raising and Aerating

Physical Methods	Mechanical Methods
• Physical raising methods such as air, water vapour or steam help products to have a light, open texture. • Recipes that need to be light have ingredients that function as raising agents such as water, milk or egg whites.	• Food preparation methods such as sieving, whisking or beating can be used to trap air. • Combinations of physical and mechanical methods work well in food preparation to make mixtures light, e.g. batters for Yorkshire puddings.

Air, Steam and Foam as Raising Agents

- **Air** is a very effective raising agent because it expands when it is heated. Air pockets swell and volume increases.
- Food preparation techniques help prevent loss of air, e.g. folding in flour when making a whisked sponge cake.
- **Steam** is produced from water in a mixture; this is a physical change.
- Steam produces light, open and uneven textures and adds volume during cooking, e.g. profiteroles.
- Moist mixtures produce steam during cooking.
- **Foams** - whisking helps trap air, creating foam.
- Ingredients containing protein form foams, e.g. milk froth, egg whites.
- Egg whites stretch and unravel to trap air to form a gas-in-liquid foam.
- Sugar stabilises foam, e.g. cold-set soufflé.
- Egg white foams set mousses.
- Cooking stabilises foam, e.g. roulade, meringue.

Chemical Raising Agents

- Chemical raising agents produce carbon dioxide when heated with a liquid.
- They cause effervescent fizzing and bubbles of gas.
- Chemical raising agents must be carefully measured.

Bicarbonate of Soda

- Bicarbonate of soda is an alkaline powder.
- It can leave a soapy aftertaste but strong flavours, e.g. gingerbread, will mask the aftertaste.
- It works more effectively with an acid ingredient such as buttermilk or cream of tartar, e.g. soda bread.
- The acid neutralises the alkali and prevents soapy aftertaste.
- Cream of tartar is an acid raising agent, which is frequently used alongside bicarbonate of soda, e.g. in scones.

Baking Powder

- Baking powder is a ready-to-use mixture of cream of tartar plus bicarbonate of soda and rice flour.

Self Raising Flour

- Self raising flour is plain flour and baking powder added together to create rise. Plain flour alone does not contain a raising agent.
- Self-raising flour can be brown or white.
- Self-raising flour contains a pre-sieved precisely measured amount of baking powder for ease and speed of use.

Gingerbread

Biological Raising Agents

- **Yeast** is a biological raising agent. It ferments to give off carbon dioxide gas.
- Fermentation in yeast is a biological (also known as biotechnological) raising agent.
- The conditions for yeast fermentation are warm temperature 25°C–35°C; moisture; food; time.
- Temperatures above 60°C during baking will inactivate and finally destroy yeast cells.
- Boiling liquids will inactivate yeast, preventing fermentation from taking place.
- Yeast is the raising agent in bread, bread rolls, buns and rich pastries (Danish pastries).
- Leavened bread contains raising agent in the form of yeast or bicarbonate of soda.
- Unleavened bread contains no raising agent and is flat in structure.

 Quick Test

1. What is the raising agent in a whisked sponge cake?
2. What happens when air is heated?
3. How does egg white trap air?
4. How can water help make a mixture light during cooking?

 Key Words

physical raising methods
chemical raising agents
yeast

Review Questions

Protein and Fat

1 Circle three foods that contain High Biological Value (HBV) protein. [3]

bread   chicken   grapes   butter

milk   broccoli   soya   lentils

2 What is it called when Low Biological Value (LBV) protein foods are eaten together to provide a good supply of all the essential amino acids? [1]

3 Over-consumption of saturated fat foods is likely to raise what level in the body? [1]

Carbohydrate

1 Unscramble the following starchy carbohydrate foods. [3]

astap ....

ototpa ....

otsa ....

2 Anyone suffering from constipation needs to eat more foods high in dietary fibre/non starch polysaccaride (NSP). Underline four foods that are high in NSP. [4]

cornflakes   lentils   cabbage   cream crackers

eggs   apples   shredded wheat   milk

3 Which type of carbohydrate is linked to tooth decay? [1]

Vitamins

1 ⊗ Circle three foods that are a good source of Vitamin C (ascorbic acid).　　　　[3]

　　cheese　　　**sardines**　　　**Kiwi fruit**　　　**peanut butter**

　　　　lemons　　　**margarine**　　　**cabbage**

2 Explain why a pregnant woman needs an increased intake of folate (folic acid).　　　[1]

....

....

3 Give **two** reasons why raw vegetables contain more vitamins than cooked vegetables.　　[2]

....

....

....

4 What is the alternative name for Vitamin C?　　　　[1]

....

Minerals and Water

1 Fill in the missing words.　　　　[3]

Iron supports the production of .... in ....

.... cells; this transports .... around the body.

2 Why is iron deficiency anaemia common in teenage girls?　　　　[1]

....

3 Iodine is a component of thyroid hormones. What do these hormones regulate
in the body?　　　　[1]

....

4 Give **two** reasons why someone may need to increase their water intake. [2]

Making Informed Choices

1 Which of the following foods would you serve to a teenage girl to increase her intake of iron? Circle **three** foods. [3]

oranges	chocolate	digestive biscuits
beef	apricots	salad

2 What is the Eatwell Guide? Tick (✓) **one** answer. [1]

a) A chart that shows you how long foods can be kept for. ☐

b) A guide that informs you of where to shop for the healthiest food. ☐

c) A guide that shows the proportions of food groups that should be eaten daily for a well-balanced diet. ☐

d) A guide that tells you how long foods have to be reheated. ☐

3 Fill in the missing words. [2]

Babies are born with a supply of i_____ stored in the l_____.

Diet, Nutrition and Health

1 What is obesity? Tick (✓) **one** answer. [1]

a) abnormal accumulation of body fat ☐

b) lack of thyroid hormone ☐

c) weakened teeth and gums ☐

d) narrowing of the arteries ☐

2 Which of the following foods would you serve to an elderly person to keep their bones as healthy as possible? Circle **three** foods. [3]

oranges   milky drinks   digestive biscuits

sausages   cheese   yoghurts   salad

3 How does someone with diabetes manage the condition? [2]

....

....

....

4 What is a healthy BMI for adults? Tick (✓) **one** answer. [1]

a) 7.5–12 ☐

b) 13.5–18 ☐

c) 18.5–25 ☐

d) 32–38.5 ☐

5 What is osteoporosis? [2]

....

....

....

Practice Questions

Cooking of Food, Heat Transfer and Selecting Appropriate Cooking Methods

1. Show you understand the key reasons for cooking food. Complete the following statements using the words in the boxes below.

shelf life	digest	variety	microorganisms	keep quality

Food is cooked to:

a) make it safe to eat by destroying .... [1]

b) help extend .... and .... [2]

c) improve .... in the diet. [1]

d) make it easier to .... [1]

2. Name **three** methods of heat transfer. [3]

....

3. Listed below are the three classifications of cooking methods. Name **two** examples of each classification. [6]

	Water-based Methods	Dry Cooking Methods	Fat-based Methods
Example 1			
Example 2			

Proteins and Enzymic Browning

1. Protein foods can be denatured in four ways. What are the **four** ways? [4]

....

....

....

2 During the cooking of a quiche the egg filling changes texture.

    a) In what way does the texture of the quiche filling change during baking?　[1]

    ....

    b) Why does the texture of the quiche filling change during baking?　[2]

    ....

    ....

3 **a)** What is the name of the protein in wheat flour?　[1]

    ....

    b) In what way is the protein content of strong flour different to plain cake flour?　[1]

    ....

4 **a)** State **two** actions that are required for wheat flour to form gluten.　[2]

    ....

    ....

    b) Gluten is formed from two proteins in wheat; glutenin is one, what is the other protein?　[1]

    ....

5 **a)** What is the term used to explain browning in foods not as a result of either dextrinisation or caramelisation?　[1]

    ....

    b) Explain one way to prevent enzymic browning when using apples.　[1]

    ....

Carbohydrates

1 Circle the correct words in the following passage. [5]

The function of starch in thickening a **solid / liquid** is known as **gelatinisation / caramelisation**. For a sauce to thicken it needs to be **chilled / heated** and also **stirred / sieved** to ensure a smooth sauce. A sauce should be heated to **setting point / boiling point** to prevent it tasting raw.

2 **a)** As bread is toasted dextrins are formed. How does dextrin affect the colour, flavour and texture of toast slices? [3]

Colour _____

Flavour _____

Texture _____

b) What is the scientific term for the effect of dry heat on starch? [1]

_____

c) Dextrins attract moisture from the air. What is the scientific term for this? [1]

_____

d) How does this process affect the quality of the toast? [1]

_____

Fats and Oils

1 Which of these statements is **not** true? Tick (✓) one answer. [1]

a) Shortening is a process that creates a hard texture in products. ☐

b) Shortening is seen in shortcrust pastry, shortbread biscuits, cookies and rich pastries. ☐

c) The process in food preparation most likely to bring about shortening is the rubbing in method. ☐

2 Fats are used in food preparation to promote characteristic textures, flavours and colours. Identify the functional properties of fat in the preparation processes on the following page. [6]

Process	Function of Process	End Product Characteristic
a) Margarine and sugar beaten (creamed) together in a mixture	Explain **two** functions	Give **two** characteristics
b) Margarine or butter rubbed into a flour mixture	**One** function	**One** characteristic

Raising Agents

1 Describe **three** functions of raising agents in food preparation. [3]

1 .... 2 .... 3 ....

2 A baker wants her shop assistants to understand raising agents. She uses examples from her shop.

Example 1 is a whisked sponge flan.

a) What is the raising agent in the flan sponge? [1]

....

b) Describe how the raising agent is incorporated into the sponge. [1]

....

....

Example 2 is a cheese scone.

c) Which raising agent is used in scones? [1]

....

d) What gas would the raising agent produce? [1]

....

e) Explain how the raising agent works during baking. [2]

....

....

Microorganisms, Enzymes and Food Spoilage

You must be able to:

- Know the growth conditions for microorganisms and enzymes and the control of food spoilage
- Know and understand that bacteria, yeasts and moulds are microorganisms
- Explain that enzymes are biological catalysts usually made from protein.

Bacteria

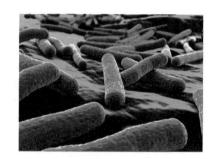

- **Bacteria** are single-celled organisms that are able to reproduce rapidly. They are also called **microorganisms.**
- Bacteria are sometimes useful and are used in cheese-making and in yogurt.
- They are sometimes harmful (pathogenic) and can cause food poisoning.

Conditions for Growth

Temperature	• Bacteria generally multiply between 5°C and 63°C. • The average ideal temperature for rapid bacterial multiplication is 37°C, which is body temperature.
Moisture	• Bacteria need moisture to live and multiply.
Time	• Under optimum conditions, bacteria will multiply every 10–20 minutes, so within seven hours one bacterium can become one million. • To control bacteria multiplying: – eat food as soon as possible after making/cooking – if food is not being eaten straight away, cool down as quickly as possible (within 90 minutes is recommended) and store in the fridge or freezer. A blast chiller will cool food more quickly than the recommended 90 minutes.
Nutrients	• Bacteria can multiply to large numbers on high-risk foods such as meat, poultry, fish, eggs and milk. These are all high-risk foods.
pH level	• Bacteria grow best at a neutral pH level, between 6.6 and 7.5. • Bacteria are unable to survive below pH 4.5. • Vinegar (acetic acid) has a pH of 3.5.

Yeast

- Yeast are single-celled plants found in the air and on skins of fruit.
- Yeast spoils the taste of food but is not harmful.
- It grows only on sugary foods, but not in sugar concentrations above 50% (jams should be made with 60% sugar).
- It can survive without air.
- It can't grow at low temperatures or survive in vinegar.
- It is destroyed at temperatures above 70 °C.

Key Point

Bacteria are found everywhere and need the right temperature, warmth, time, nutrients, pH level and oxygen to grow and multiply.

- Yeast is a very helpful organism. It is used in baking bread, where carbon dioxide is used as a raising agent, and in wine making and brewing.

Moulds

- **Moulds** are a type of fungus, that settle on food and grow into a visible plant.
- Moulds grow on many foods, such as bread, cheese and meat.
- They like slightly acid conditions.
- They need moisture and warmth (20 °C–40 °C), but are destroyed by heat <70 °C.
- Moulds can survive in the fridge but not in the freezer.
- Mould on food is a sign that it is not very fresh or has been stored incorrectly. Some mould can result in allergic reaction and respiratory difficulties.

Enzymes

- **Enzymes** in food can be a problem for food storage.
- The cells break open, the enzymes escape and react with other parts of the food. Soft spots appear on fruit and vegetables and makes meat smell and taste bad.
- Denaturing the enzymes can help to preserve the food, e.g. through heat, use of acids, strong alkalis or salt.
- Enzymes are chemical catalysts that are found in all cells.
- Enzymes break down plant and animal tissues, causing fruit to ripen, meat to tenderise and enzymic browning (also known as oxidation) to speed up.

Ways to Prevent Oxidation (Enzymic Browning)

- Adding lemon juice (an acid) to a fruit salad prevents browning.
- Blanching vegetables before freezing prevents discolouration.
- Removing air by immersing potatoes in water will prevent browning.
- Refrigeration or freezing will slow down browning.
- The removal of moisture (dehydration) will prevent browning, however it is a slow process and the browning reaction is quick.

Quick Test

1. What are microorganisms?
2. What is the ideal temperature for bacterial growth?
3. How can we control bacterial growth?
4. Name **two** food products that use yeast.

Key Words

bacteria
microorganisms
moulds
enzymes

Microorganisms in Food Production

You must be able to:

- Demonstrate knowledge and understanding of the use of microorganisms in food production, including moulds in the production of blue cheese, yeasts as a raising agent in bread and the use of bacteria in yogurt production.

The Dairy Industry

- It would be impossible to make cheese without a **starter culture**.
- As the culture grows in the milk, it converts the sugar lactose into lactic acid, which ensures the correct level of acidity and gives the cheese its moisture.
- As the cheese ripens, the culture gives it a balanced aroma, taste, and texture.
- Choosing the right mixture of culture is essential for a high-quality cheese.

Cheese Production and Moulds

- Cheeses that rely on moulds for their characteristic properties include:
  - blue cheese
  - soft ripened cheese (such as camembert and brie)
  - rind-washed cheese (such as époisses and taleggio).

Blue Cheese

- To make blue cheese:
  - the cheese is treated with a mould
  - as the cheese matures, the mould grows
  - this creates blue veins within the cheese, giving the cheese its characteristic flavour, e.g. stilton and roquefort.

Blue cheese

Soft Ripened Cheese

- To make soft ripened cheese:
  - P. camemberti is allowed to grow on the outside of the cheese, causing the cheese to age from the outside in, forming a soft white crust and runny inside e.g. brie and camembert.

Soft ripened cheese

Rind-washed Cheese

- To make rind-washed cheeses:
  - rind-washed cheeses also ripen inwards but they are washed with brine and other ingredients, e.g. beer and wine, which contain mould
  - this makes them attractive to bacteria, which adds to the flavour, e.g. limburger.

Rind-washed cheese

Yoghurt

- In yoghurt, the culture is responsible for the taste and texture of the final product.
- In recent years **probiotic** cultures have become popular in dairy products because of their health benefits.
- Probiotic cultures are carefully selected strains, and there is good evidence that they help improve digestion, safeguard the immune system, and keep the body's intestinal flora in balance.
- Probiotic cultures are classified as a functional food.

The Meat Industry

- Meat starter cultures are used to make dried, fermented products such as salami, pepperoni, chorizo and dried ham.
- Lactic bacteria develop the flavour and colour of the products.
- A wide variety of moulds are used to ripen the surface of sausages, preserving the natural quality of the product and controlling the development of flavour.

Key Point

Microorganisms (bacteria) are used to make a wide range of food products. Bacteria are used to make cheese, yogurt, and bread. The most important bacteria in food manufacturing are the Lactobacillus species.

Yeast

- Yeast is used in bread-making as well as in making beer and wine.
- Yeast is a microorganism.
- Yeast requires sugar to grow.
- In bread making, yeast will:
  - leaven the dough by producing CO_2
  - through fermentation and its enzymic action on other ingredients, create a stretchy dough
  - contribute to the flavour of the bread.

Quick Test

1. What is the most important bacteria used in food manufacturing?
2. What is blue cheese treated with to give it its taste and texture?
3. In what ways are probiotic cultures considered to have health benefits?

Key Words

starter culture
probiotic

Bacterial Contamination

You must be able to:

- Know and understand the different sources of bacterial contamination
- Know and understand the main types of bacteria that cause food poisoning
- Demonstrate knowledge and understanding of the main sources and methods of control of different food poisoning bacteria types
- Recognise the general symptoms of food poisoning.

Bacteria

- Bacteria can be found everywhere, including raw food, people, air and dust, equipment and utensils, soil, pests, water and food waste.

The Dangers of Bacteria

- It is essential to control the conditions that allow bacteria to multiply and cause illness, e.g. stick to strict time and temperature controls.
- You can become ill if you eat food that is contaminated by certain bacteria (pathogens) and viruses.
- Kitchens provide the ideal conditions for bacteria growth.
- Bacteria are microscopic. You cannot tell if a food is contaminated by just looking at it.

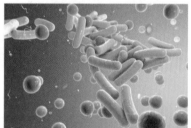

Food Poisoning Bacteria

- **Food poisoning** is caused by bacteria multiplying in or on food.

Pathogenic Bacteria	Foods Affected	Symptoms	Onset	Special Note
Salmonella	Raw meat; Eggs; Seafood; Dairy products	Diarrhoea; Vomiting; Fever	12–36 hours	May be fatal to the elderly and babies; Found in human and animal excreta
Staphylococcus Aureus	Cooked sliced meat; Dairy products; Anything touched by hand	Vomiting; Diarrhoea; Abdominal pain	1–6 hours	Present in nasal passages, throat and skin; Associated with dirty food handlers; Good personal hygiene is crucial
Clostridium perfingens	Raw and cooked meat and meat products	Nausea; Diarrhoea; Abdominal pain	8–22 hours	
Clostridium Botulinum	Incorrectly canned meat, fish or vegetables	Paralysis; Difficulty breathing; Double vision; Nausea; Vomiting	12–48 hours	Rare
Bacillus cereus	Cooked rice, pasta and cereal foods	Nausea; Vomiting; Diarrhoea	1–6 hours	

Food-borne Disease

- Food-borne diseases are caused by pathogenic microbes (bacteria and viruses) carried on food. These microbes do not multiply in the food but in the person who has eaten the food.

Pathogenic Microbes	Foods Affected	Symptoms	Onset	Special Note
Escherichia-Coli (E-Coli)	Raw meat; Untreated milk and water	Vomiting; Blood in diarrhoea; Kidney damage or failure	12–24 hours	Causes gastro-enteritis in humans
Listeria Monocytogenes	Soft cheeses; Pate; Unpasteurized milk; Undercooked meat; Incorrectly heated cook-chill meals	A range of symptoms from mild flu-like symptoms to septicemia, meningitis and pneumonia	No specific time	Can cause miscarriage, premature labour and birth
Campylobacter	Meat; Shellfish; Untreated water; Washing raw poultry	Diarrhoea; Headache; Fever; Abdominal pain	1–11 days	Easily transmitted between humans. Raw meat and poultry **MUST NOT** be washed as this spreads the bacteria
Norovirus	Shellfish; Raw vegetables; Salads	Nausea; Vomiting; Diarrhoea; Abdominal pain; Fever	1–2 days	Projectile vomiting The virus can survive for several days if not cleaned up properly

Preventing Contamination

- Preventing **contamination** is the key to food safety.
- Keep the kitchen clean and tidy. Clean and disinfect all areas, equipment and utensils used to prepare food.
- Keep food covered. Handle the food as little as possible.
- Store food correctly. Cook food thoroughly.
- Remove food waste and rubbish.

Food Poisoning

- Food poisoning is caused by harmful bacteria multiplying in or on food.
- Symptoms usually pass in a few days and the sufferer will make a full recovery.
- Food poisoning can affect anyone, but some people may suffer more than others, such as very young children and babies; elderly people; those who have had a serious illness or are recovering from a serious illness; pregnant women and nursing mothers; people who have allergies.

Key Point

Bacterial contamination is the presence of harmful bacteria in our food, which can lead to food poisoning and illness. As a food handler, you must do everything possible to prevent this contamination.

Key Words

pathogens
food poisoning
contamination

Buying and Storing Food

You must be able to:

- Know and understand the food safety principles when buying and storing food
- Know and understand temperature control and ambient storage
- Understand danger zone temperatures and be able to describe the correct uses of domestic fridges and freezers
- Understand date marks.

Buying Safe Food

- Make sure the food is safe for consumption.
- Check for signs of damage to tins, packets and packaging.
- Check the date marks – the **use-by date** and **best-before date**.
- Stock rotation makes sure that food is used within date (when it is safe) and prevents unnecessary waste.

Use-by Dates

- A use-by date is a safety date found on foods and their packaging.
- Use-by dates are found on highly perishable, packaged food such as meat, fish and dairy products that require chilling and have a short shelf life.
- It is an offence for businesses to sell or use food that has passed its use-by date.

Best-before Dates

- Best-before dates usually appear on less perishable foods that have a long shelf life, such as canned, dried and **frozen food** products.

Storing Food

- It is important to store food properly.
- Food should be wrapped, covered or kept in a suitable clean container.
- Raw meat, poultry, fish and eggs should always be stored away from ready-to-eat foods. Raw vegetables, including salad leaves, may contain pathogens from the soil and should be kept away from ready-to-eat/high-risk foods.
- The refrigerator should be cleaned regularly.
- Dry food ingredients, e.g. dried pasta, tinned tomatoes, herbs and spices should be stored in cupboards that are clean and dry.
- Frozen food still could have bacteria present but they are dormant. Enzymes that cause food spoilage are slowed but not stopped unless inactivated – this is the reason for blanching vegetables before freezing.

- Food ingredients and finished products must be:
  - at the correct temperature
  - protected from contamination
- Food ingredients should be used according to the date marks on the food or its packaging.
- Using food within the date marks avoids waste and reduces the risk of cross contamination.

Chilled Food Storage

- Fridges should be set to operate at a temperature between 1 °C and 4 °C to make sure that **chilled food** is held below 5 °C.
- Door handles should be cleaned and disinfected.
- Raw ready-to-eat food and high-risk foods should be kept above raw food (for cooking) in the fridge.
- Food must be kept covered, and food-safe containers and food labels used.
- Never put warm food in a fridge, it will cause the temperature in the fridge to rise into the danger zone (5 °C–63 °C).

Frozen Food Storage

- Bacteria are dormant in the freezer.
- Frozen food should be stored at –18 °C or below.
- Food should be well wrapped to prevent freezer burn through loss of moisture.
- Food bought as frozen should be stored in the freezer as soon as possible.

Defrosting

- If frozen foods are not defrosted thoroughly, ice crystals remain at the centre. Cooking will melt the ice but the correct core temperature may not be achieved and bacteria may survive.
- Protect defrosting food from being contaminated.
- If high-risk food is defrosted at room temperature, bacteria will start to multiply on the surface of the food.
- Defrosted food must be treated like chilled food.

 Key Point

When handling food at any stage, from buying the food through to correct storage, steps must be taken to prevent contamination. Everything possible must be done to control the conditions that allow bacteria to multiply, causing food poisoning.

Quick Test

1. What are the **two** date marks you need to check when buying food?
2. What is the recommended temperature for chilled food?
3. It is acceptable to put warm food in the fridge. True or false?
4. What is the recommended temperature for a freezer?
5. What happens to food that is not defrosted thoroughly?

 Key Words

use-by date
best-before date
frozen food
chilled food

Preparing and Cooking Food

You must be able to:

- Know and understand the food safety principles when preparing and cooking food, including the importance of personal hygiene and clean work surfaces
- Understand about cross contamination and temperature controls
- Know the factors surrounding high-risk food
- Demonstrate knowledge of the correct use of temperature probes.

Personal Hygiene

- Food handlers are a common source of pathogenic bacteria.
- When preparing food, you must maintain the highest possible standards of personal **hygiene** in order to avoid contaminating food and causing illness.
- A clean apron should always be worn to protect the food from contamination from your clothes.
- All jewellery should be removed when cooking, food handlers must tie back long hair or wear a hairnet/hat, and fingernails should be kept short and clean with no nail varnish.
- Licking your fingers or utensils when cooking must be avoided.
- Wash your hands:
  - before preparing food
  - between handling raw and **high-risk foods** or ready-to-eat foods
  - after you have been to the toilet
  - after sneezing/coughing
  - after changing a waterproof plaster
  - after cleaning, after handling food waste or rubbish and known allergens.

Preparing Food

- Food needs to be protected from contamination during preparation, as the food is usually uncovered while being handled.
- Food should be handled as little as possible.
- Food must be kept out of the **danger zone** (5 °C–63 °C) as bacteria will multiply.
- To prevent cross contamination, raw food and cooked food must be prepared separately using separate utensils (different chopping boards and knives). Red chopping boards and knives are used for raw meat and should only be used for this purpose.

Preparing High-risk Foods

- When preparing food, some foods are more of a risk than others: these are high-risk foods.
- Protein-based foods, moist foods and ready-to-eat foods that require no further cooking are high-risk foods.
- High-risk foods require strict time and temperature controls.

Key Points

When preparing and cooking food it is important to have high personal and food hygiene standards.

High-risk foods include:
- cold cooked meat
- poultry
- fish and seafood
- ready meals containing gravy or sauce
- egg products, such as quiche and homemade mayonnaise
- although not high in protein, cooked rice, pasta and cereals are classed as high-risk as they can provide a moist environment where certain bacteria can multiply
- dairy-based products, e.g. cream cakes and desserts.

Cooking and Reheating

- Pathogenic bacteria can be destroyed and food made safe by thorough cooking. If food is not cooked correctly, pathogenic bacteria may survive and cause food poisoning.
- To make sure bacteria are killed, the thickest part of the food (the core) should reach 70 °C for two minutes.
- Liquids should be stirred to avoid cold spots.
- Large items should be turned to ensure even cooking.
- Poultry, pork and minced meat products must be thoroughly cooked to kill the bacteria.
- If a microwave oven is to be used, always follow the manufacturer's guidelines.
- Hot holding is the process of storing food warm for service. Food must be kept at a temperature of no lower than 63 °C for a maximum of 90 minutes before being discarded.
- Food must not be **reheated** more than once.
- Bacteria will multiply during repeated cooling and reheating.
- Reheated food should reach a **core temperature** of 75 °C.

Key Point

Food must be cooked thoroughly to make it safe to eat.

Cleaning

- All food contact surfaces and hand contact surfaces must be cleaned and disinfected after use. Best practice is to 'clean as you go'.
- Food preparation areas must be kept clean to protect food from: bacterial contamination, physical contamination and chemical contamination.
- Food preparation areas must be kept clean to prevent slips and trips.

Types of Food Poisoning

- Food poisoning can occur through:
  - Microbiological contamination – this is the germ itself contaminating the food.
  - Physical contamination – this is when items physically drop into food, i.e. rings, nails, hair.
  - Chemical contamination – this is when chemicals get into food, i.e. cleaning chemicals.

Temperature Probes

- Temperature probes are used to take the core temperature of food.
- Care must be taken to ensure the probe does not cause cross contamination: it must be cleaned and disinfected after every use.

Quick Test

1. List **four** occasions during food preparation when you must wash your hands.
2. What temperature control and time during cooking kills most bacteria?

Key Words

hygiene
high-risk foods
danger zone
reheating
core temperature

Review Questions

Cooking of Food, Heat Transfer and Selecting Appropriate Cooking Methods

1 Improving palatability is one of the main functions of food preparation and cooking. Fill in the blanks to complete the following text. [5]

Palatability means how acceptable the sensory characteristics of a food are, e.g. how

appetising it is. Making food palatable develops ....

improves .... and .... , reduces

.... and changes ....

2 Many methods of cooking use a combination of heat transfer. Explain how methods of heat transfer can combine when boiling potatoes. [6]

....

....

....

....

....

....

....

....

Proteins and Enzymic Browning

1 What does protein denaturation mean? Give **one** example, and explain this example (continued on page 69).

Meaning [2]

....

....

Example [2]

_____

_____

2 Choose the correct words from the boxes below to complete the sentences. [5]

| elastic | kneading | short | gluten | structure |

Wheat flour contains the protein _____ which makes a dough

stretchy and _____. Different cooking techniques can affect gluten

formation, e.g. _____ helps increase gluten formation in

breadmaking, whereas rubbing fat into flour makes _____ gluten

strands to aid the texture. Gluten forms the _____ of pastries,

breads and cakes.

Carbohydrates

1 What are **two** changes that can be noticed when a solution of sugar and water is heated? [2]

_____

_____

2 Complete the boxes to give the scientific terms for changes that occur during baking. [2]

Ingredients	Effect of Baking	Scientific Term	Characteristic
Flour	Starch browns		Golden colour
Sugar	Sugars brown		Golden colour

3 Give **one** practical application of each of the scientific terms that complete the table in question 2. [2]

1 _____ 2 _____

4 A chef is planning to use a roux sauce to coat steamed cauliflower. Which recipe would be his best choice and why?

Recipe A	Recipe B
500 ml milk	500 ml milk
10 g flour	25 g flour
10 g butter	25 g butter

a) Choice .... [1]

b) Explain your answer. [4]

....

....

Fats and Oils

1 A student decides to make shortcrust pastry by hand.

a) What is the name of the method he would use? .... [1]

b) Describe how the process makes a short textured pastry. [6]

....

....

....

....

....

2 Choose the correct words from the boxes below to complete the following sentences (continued on page 71). [6]

(continued on page 71)

natural	emulsions	lecithin	immiscible	stable	separates

.... are mixtures of liquids that do not normally mix together.

Vinaigrette dressing is an example of an .... liquid, it

.... on standing.

Mayonnaise is an example of a _____ emulsion. The

_____ in egg yolks helps make emulsions stable because it is a

_____ emulsifier.

3 Indicate which of the following statements are true and which are false. Circle the
correct answer. [4]

a) The plasticity of a fat enables fat to perform effectively when using the rubbing-in method.
Circle the answer.

True   False

b) Plasticity means a fat cannot change properties. Circle the answer.

True   False

c) Temperature is an important factor in the plasticity of fats. Circle the answer.

True   False

d) Plasticity helps a fat to be spreadable. Circle the answer.

True   False

4 A student is using the creaming method to make some small cakes.

a) Which **two** ingredients would be creamed together first? [2]

_____

b) During the preparation of the cake he needs to check the mixture. State **one**
characteristic of the creamed mixture he might look for. [1]

_____

c) Explain why the mixture would show this characteristic. [2]

_____

_____

d) How would this help the cakes during baking? [2]

_____

_____

Review Questions

Raising Agents

1 Explain **four** conditions needed for yeast fermentation.

1 .... [2]

2 .... [2]

3 .... [2]

4 .... [2]

2 A student has been asked to investigate raising agents to make products with light and airy textures.

a) What are raising agents? [3]

....

....

....

b) Complete the table naming **three** differing products the student could make using **three** different raising agents. [6]

Name of Product	Raising Agent Used

3 During food preparation and cooking, physical and mechanical methods are used to help make mixtures light. What equipment or method can be used to assist aeration and lightness when using the following ingredients?

a) Flour for baked goods .... [1]

b) Eggs and caster sugar in a whisked sponge .... [1]

c) What is the main effect of using these methods? .... [1]

Microorganisms, Enzymes and Food Spoilage

1 What is an enzyme? [1]

....

2 Some fruits and vegetables go brown when they are cut and exposed to oxygen. What is the scientific term for when fruits and vegetables go brown? [1]

....

3 Name **one** fruit that could turn brown. [1]

....

4 Suggest **one** method that can be used to slow down the browning process. [1]

....

....

Microorganisms in Food Production

1 Microorganisms are used in the production of: [1]

a) milk, butter and cheese. ☐

b) bread, cheese and yogurt. ☐

c) salami, sausages and burgers. ☐

d) bread, butter and oil. ☐

2 Which microorganism is used to make salami, pepperoni and chorizo? [1]

....

3 Yeast is most active at cold and high temperatures. Is this true or false? Tick (✓) **one** answer. [1]

True ☐ False ☐

Practice Questions

Bacterial Contamination

1 Bacterial contamination causes food poisoning. Is this true or false? Tick (✓) one answer. [1]

True ☐   False ☐

2 Which one of the following is important in helping prevent bacterial contamination? [1]

a) Clean as you go. ☐

b) Make sure all the kitchen lights are working. ☐

c) Check all equipment is working. ☐

d) Store heavy equipment at a lower level. ☐

3 Explain what the 'use by' date means on a food product. [1]

....

....

Buying and Storing Food

1 Refrigeration is used for food storage to slow down spoilage. Is this true or false?
Tick (✓) **one** answer. [1]

True ☐   False ☐

2 Complete the following sentences using the words in the boxes (continued on page 75).

perishable	1 °C and 4 °C, below 5 °C
cooked	−18 °C and below

a) When preparing food, keep raw and .... foods separate. [1]

b) Chilling high risk/.. foods will slow down the growth of
bacteria. [1]

c) Check the internal temperature of a refrigerator to make sure it is between

.... [1]

d) The temperature of your freezer should be .... [1]

Preparing and Cooking Food

1 Which of the following is most likely to cause cross contamination? Tick (✓) **one** answer. [1]

a) Using ready to eat foods within their use-by-date. ☐

b) Placing ready-to-eat foods above raw foods in a fridge. ☐

c) Using the same knife to cut raw chicken and cooked ham. ☐

d) Storing raw chicken in a covered container at the bottom of the fridge. ☐

2 Why should leftover food only be reheated once? [1]

....

....

....

3 Circle the foods that are considered to be high risk. [2]

cooked rice   potatoes   chocolate

tomatoes   cooked chicken   bananas

apples   biscuits

Food Choices

You must be able to:

- Understand that religions, customs and beliefs influence food choice
- Know about conditions that may be caused by an intolerance or allergy to food.

Religious Diets

- Some religions have their own dietary laws and rules. For example:
  - Jewish people eat **kosher** food, which means food that has been prepared according to Jewish law.
  - Muslims will only eat meat that is **halal**, where animals are slaughtered in a religiously approved way.

Religion	Dietary Requirements
Judaism	• No shellfish or pork • Only kosher meat • No dairy foods are eaten with meat in a meal
Hinduism	• No beef or beef products • Many Hindus are vegetarian
Islam	• No pork • Only halal meat can be eaten
Sikhism	• No beef • Many Sikhs are vegetarian or ovo-lacto vegetarian
Christianity	• No particular dietary requirements
Buddhism	• Vegetarian
Rastafarianism	• Vegetarian or vegan • White fish sometimes eaten (but no shellfish)

Vegetarians

- Vegetarians don't eat meat, poultry, fish, or products such as gelatin that have been obtained by killing animals.
- Ovo-lacto vegetarians eat eggs and dairy products (but only cheese made with vegetable rennet).
- **Lacto vegetarians** eat dairy products and honey, but not eggs.
- Vegans do not eat any food with an animal origin. To make sure that their diet is not deficient in certain nutrients (e.g. iron and vitamins D and B12, which are commonly found in animal products), vegans can get the necessary nutrients in their diet from a variety of sources including peas, beans, lentils, nuts and seeds. Soya and fortified products are also sources of protein.
- The reasons why people become vegetarian include: religious dietary laws; **ethical** reasons, e.g. meat and fish farming being wasteful of the Earth's resources; moral reasons, e.g. animal cruelty; health reasons, e.g. allergies; because they come from a vegetarian family.

Key Point

A diet consists of the foods that a person chooses to eat.

Medical Conditions

- **Diabetes** is a condition caused because the pancreas doesn't produce any, or enough, insulin to control the amount of sugar in the blood.
- **Type 1 diabetes** is often diagnosed in childhood and is not associated with excess body weight.
  - It is treated with insulin injections or using an insulin pump.
  - It can't be controlled without taking insulin.
- **Type 2 diabetes** is usually diagnosed in over 40-year-olds and is often associated with excess body weight, high blood pressure and/or cholesterol levels at diagnosis.
  - It is treated initially with medication/with tablets. It is sometimes possible to come off diabetes medication.
  - 90% of people with diabetes have Type 2 diabetes, which is best treated with a healthy diet and increased physical activity.

Allergies and Intolerances

- **Coeliac** disease is a condition where people have an adverse reaction to gluten, a protein found in wheat, barley, rye, and also oats, which contain a similar substance to gluten.
- Xanthan gum is added to gluten-free flour in order to make the product elastic and stretchy, which can be reduced by the absence of gluten.
- Coeliacs cannot absorb nutrients if they eat gluten. This causes severe pain and can lead to anaemia and **malnutrition**.
- **Lactose intolerance** is caused when the body is unable to digest lactose (a sugar found in milk and dairy products). Lactose intolerance causes stomach upset.
- An **allergy** to nuts can cause **anaphylaxis**, a reaction that can be fatal. People with severe allergies carry an epi-pen in case of an attack.
- Food products must be labelled if they contain nuts.
- People with a nut allergy have to check that food does not include nuts as an ingredient and that food has been produced in a nut-free environment.

Products displaying the Crossed Grain symbol have been licensed by Coeliac UK and are safe to eat for those following a gluten-free diet.

Key Point
If you can't tolerate certain foods, you have to change your diet.

GLUTEN FREE   DAIRY FREE   SOY FREE   EGG FREE   NUT FREE   PEANUT FREE   CORN FREE   NO TRANS FAT

NO SUGAR ADDED   NO SUGAR   VEGETARIAN   ORGANIC

Key Words

kosher
halal
vegetarian
ovo-lacto vegetarian
vegan
lacto vegetarian
ethical
diabetes
coeliac
malnutrition
lactose intolerance
allergy
anaphylaxis

Quick Test

1. Which religions traditionally do not eat pork?
2. Which type of vegetarian would not eat honey?
3. Which foods can people with coeliac disease not include in their diet?

British and International Cuisines

You must be able to:

- Understand the meaning of 'cuisine' in terms of the foods related to the traditional eating habits of certain countries
- Learn about the cuisine of two other countries as well as British traditional cuisine.

Traditional British Food

- British food makes use of ingredients produced in the local area.
- British cheeses originate from different parts of the country, e.g. Cheddar (Somerset), Wensleydale (Yorkshire); Red Leicester, Double Gloucester.
- Each cheese has its own distinctive colour, flavour and texture and is made using ingredients from its region of origin.
- There are lots of **regional** dishes, including Cornish pasties, Lancashire hot pots, Melton Mowbray pies, Eccles cakes, fish and chips and suet-based puddings such as Sussex Pond Pudding.
- Regional dishes normally have historic links. For example, Cornish pasties were eaten by tin miners working underground. The crimped edge was the handle for dirty hands so that the rest of the pastry could be eaten quickly and without mess.

Traditional British meal

Modern British Food

- We now live in a country that is **multicultural** and people travel to holiday destinations worldwide and are exposed to the **cuisines** of many other countries. Cheaper air travel prices mean that these countries are more easily accessible to a greater number of people than ever before.
- Supermarkets and specialist shops provide a vast range of ingredients, such as herbs, spices, fruit and vegetables from global cuisines.
- Because of influences from other countries, meals in the UK now contain a wide variety of foods.
- Recipes have been adapted to meet our tastes, e.g. Chicken Tikka Massala (a rich, flavoursome chicken dish without chilli) was invented in the UK but is probably based on Bangladeshi cuisine.

Cornish pasty

Other Cuisines

Spain: Tapas and Paella

- Tapas consist of a wide variety of appetisers or snacks served on small plates and chosen from a menu list.

- Tapas may be cold, e.g. mixed olives or tortilla (a potato omelette) or hot, e.g. chopitos (battered and fried baby squid), garlic prawns (gambas pil pil) and patatas bravas (cubes of potato in a spicy tomato sauce).
- Paella is widely served in restaurants and is traditionally eaten by large groups of people in the street during fiesta times.
- Paella is based on rice from Valencia (a region of Spain) and cooked in a wide flat pan with a mixture of locally sourced foods such as seafood, meat, vegetables and spices, e.g. saffron.

Japan: Fish, Noodles and Rice

- Fish is common in traditional Japanese cuisine as most of the population live near a coastline.
- Other seafoods such as seaweed are also important in the Japanese diet as it is rich in protein and has a flavour that many people love.
- As well as rice, udon noodles are an important ingredient.
- A typical Japanese meal consists of a bowl of rice (**gohan**), a bowl of miso soup (**miso shiru**), pickled vegetables (**tsukemono**) and fish or meat.
- Sashimi consists of thin slices of raw fish pH cooked or other seafood served with spicy Japanese horseradish (**wasabi**) and soy sauce (shoyu).
- Sushi consists of seafood, vegetables and egg served on vinegared rice.

> **Key Point**

‘Cuisine’ relates to the established range of dishes and foods of a particular country or region.

> **Key Point**

‘Cuisine’ is also concerned with the use of distinctive ingredients and specific cooking and serving techniques.

> **Quick Test**

1. Name **two** British cheeses that are not mentioned on this spread.
2. How well does the traditional British meal match the Eatwell Guide?
3. Why does a Cornish pasty have a crimped edge?

> **Key Words**

regional
multicultural
cuisine

Sensory Evaluation

You must be able to:

- Understand how to taste food products, using your senses, accurately
- Know about a range of sensory testing methods.

Tasting Food and Drink

- There are five **senses** that are used to **taste** food and drink. A combination of these senses helps you decide if you like a food.
- **Taste:** the tongue can detect five basic tastes: bitter, sweet, salt, umami – a savoury taste, acid/sour
- **Sight:** food's appearance influences how much we want to eat it.
- **Smell:** the **aroma** of food reaches the nose before it reaches the mouth and is tasted.
- **Touch:** what food feels like in the mouth (**texture**).
- **Hearing:** what food sounds like, e.g. sizzling.
- The senses help to develop personal food preferences (likes/dislikes) and evaluate foods, either through preference or discrimination tests.
- A range of accurate sensory words should be used when describing food. These usually come under the headings of: appearance, flavour, texture and aroma.

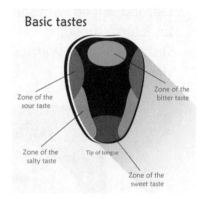

Basic tastes

Zone of the sour taste

Zone of the salty taste

Tip of tongue

Zone of the bitter taste

Zone of the sweet taste

Controlled Sensory Analysis

- Here are the stages in carrying out a controlled **sensory analysis**:
  - Invite people to be your testers in the sensory analysis.
  - Find a quiet area to work.
  - Give each tester a cup of water to cleanse their **palate** between each sample.
  - Provide small samples of food with clean spoons or forks for each sample. Provide a sheet to record results for each tester.
  - Foods should be identified using codes or symbols so that tasters are not influenced by brand names.
  - Food products should be tested carefully and results recorded accurately.

> **Key Point**
>
> The human **olfactory** system (smell) and taste sensors are important when tasting food.

Paired Preference Tests

- Testers are asked to taste two similar products.
- They are then asked which product they prefer.

Triangle Testing

- Three samples are tested but two are the same.
- The aim is to find out if the tester can pick out which sample is different, e.g. a sauce made with 15% fat or 5% fat. This

Triangle Testing

Which one is the odd one out?

test helps to work out whether a 'healthier' product can be developed without losing taste.

Ranking Tests

- People are asked to taste several products and award scores to decide on the order of preference (best to worst).
- All the samples should be coded.

Rating Tests

- People are asked to say how much they like or dislike a **sensory characteristic** of a product. This is called a **rating**.
- They use a hedonic scale to award a number (using a list provided), or they can indicate which symbol they think is best (most useful for young children).
- Sensory characteristics include sweetness, flavour, colour, texture.

Sensory Characteristic	Testers				Total
	1	2	3	4	
Evenly spread toppings					
Golden brown cheese					
Cheese aroma					
Tomato flavour					
Pizza base texture					
Overall flavour					

KEY
1 = dislike a lot
2 = dislike a little
3 = neither like nor dislike
4 = like a little
5 = like a lot

Sensory Profiles

- The results of sensory tests are often displayed visually using charts and sensory profiles, such as the star profile/radar diagram below.

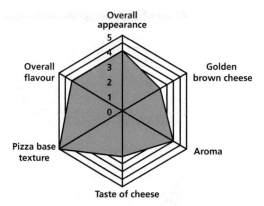

Example: sensory profile for a cheese and tomato pizza.

Key Point

Accurate sensory testing of foods helps manufacturers and cooks develop food products and improve recipes.

Key Words

senses sensory
taste analysis
aroma palate
texture sensory
olfactory characteristic
rating

Quick Test

1. What is another name for the feeling of food in mouth?
2. Where are the taste sensors in your body?
3. Why is it important to use codes/symbols when tasting foods?

Food Labelling

You must be able to:

- Know which information is legally required for a food label
- Explain how this information will help the consumer
- Understand the ways in which nutritional labelling can be presented.

Food Labels

- The information on a food label is controlled by EU regulations.
- Pre-packaged foods have information labels to inform consumers.
- The following information on food labels is required by law:
  - the name of the food
  - weight or **volume** (indicated by 'e', which stands for 'estimated weight', is normally placed after the total weight of the product in grams – this explains that the weight is estimated and allows for differences)
  - ingredients list (from largest to smallest)
  - **allergen** information
  - GM (genetically modified) ingredients
  - date mark and storage
  - cooking instructions – to ensure food is safe to eat
  - place of **origin**
  - name and address of **manufacturer** (in case of complaint)
  - lot or **batch** mark (for traceability)
  - E numbers – chemical additives that have been approved for use in the European Union
  - nutritional information (from 2016).

Recent Changes to Food Labels

- From December 2016 the rules for **nutrition labelling** in the EU and the FSA in the UK must be followed on **pre-packed** foods.
- The nutrition declaration must include the:
  - energy value in both kilojoules (kJ) and kilocalories
  - amounts in grams (g) of fat, saturates, sugars, protein and salt.
- You must have nutrition labelling if:
  - you make a nutrition or health claim
  - vitamins or minerals are added to the food.

Date Marks and Storage

- **Use by/best before** dates indicate the date when the food is safe to eat before the quality begins to deteriorate and bacterial numbers rise.
- **Display by** dates indicate shelf-life in the retail store.
- **Sell by** dates show when the product should be removed from sale to the customer.
- **Storage** dates indicate how the food should be stored in order to maintain freshness and quality.

Key Point

Make sure that you have a good working knowledge of all the information that must be present on a food product label to inform a consumer.

Key Point

EU = European Union

FSA = Food Standards Agency

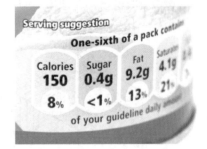

Key Point

You should know the accepted methods of displaying nutritional information on food labels.

The **name** identifies the food. Processed foods must also be identified by the cooking method, e.g. smoked, roast.

Contains GM soya or maize allows consumers to make an ethical choice.

Weight / volume: 'e' indicates approximate weight – allows for tolerances.

Manufacturer's name or address allows consumers to contact the manufacturer. **Country of origin.**

Shelf life details: the best before date indicates that food is safe to eat but the quality will begin to deteriorate. After the use by date there is a risk of food poisoning. The display by / sell by date tells the retailer what to do with the product.

Ingredients are listed in **descending order of weight**. The largest amount is first and the smallest amount is the last.

Contains nuts or **may contain traces of nuts** indicates whether the product may have been in contact with nuts during manufacture.

Instructions for use / heating instructions ensure that the product is cooked at the right temperature and that it's safe to eat.

Storage instructions are shown in words, symbols or temperatures.

The allergens that must be shown on the label are:
- celery
- cereals containing gluten
- crustaceans
- eggs
- fish
- lupin
- cow's milk
- molluscs
- mustard
- nuts
- peanuts
- sesame seeds
- soybeans
- sulphur dioxide (sulphites).

Allergen Information

- There are 14 allergen foods that must be indicated on the label if they are present in the food (see alongside).

Nutrition Labelling Methods

- A front of package label should contain:
  - Information on the energy value in kilojoules (kJ) and kilocalories (kcal) per 100g/ml and in a specified portion of the product.
  - Portion size information expressed in a way that is easily recognisable by, and meaningful to, the consumer.
  - % RI (Reference Intake) information based on the amount of each nutrient and energy value in a portion of the food.
  - Colour coding of the nutrient content of the food. Companies may additionally include the descriptors 'High', 'Medium' or 'Low' (HML) together with the colours red, amber or green respectively to reinforce their meaning.
- **Back of pack** labelling (compulsory in the EU from 2016).
  - The current rules specify the nutrients that can be included.
  - The information has to be presented per 100 g/ml but could also be provided per portion.

Each 1/2 pack serving contains

MED	LOW	MED	HIGH	MED
Calories	Sugar	Fat	Sat Fat	Salt
353	0.9g	20.3g	10.8g	1.1g
18%	1%	29%	54%	18%

of your guideline daily amount

Quick Test

1. When reading a food label, what weight of the food is used?
2. What does EU stand for?
3. What must be included in the nutrition declaration as well as fat, saturates and salt?

Key Words

volume
allergen
origin
manufacturer
batch
nutrition labelling
pre-packed

Factors Affecting Food Choice

You must be able to:

- Show that you understand that people choose what to eat for a complex variety of reasons
- Provide some reasoned suggestions for food choice based upon occasion, health, lifestyle and income.

Physical Activity Level (PAL)

- There are UK Government guidelines for 5–18 year olds regarding physical activity.
- These guidelines promote the benefits of being active for young people.
- The guidelines suggest at least 60 minutes of physical activity each day.
- Physical activity improves **cardiovascular** and bone health.
- Physical activity helps to maintain a healthy weight.

Healthy Eating

- The **Eatwell Guide** is a government guide that advises the public what and how much to eat.
- It makes healthy eating easier to understand by showing a visual image of the types and **proportions** of foods needed for a well-balanced diet. (see page 28).

Key Point

Many factors affect the food choices that people make.

Income and Cost of Food

- The income a household has influences food choices.
- Low-income households have to make difficult choices regarding healthy foods, cost and quality, e.g. protein foods and fresh fruit and vegetables are generally more expensive than starchy foods.
- Where people shop and what they buy is affected by cost.
- Some food **retailers** market their foods based on high quality and others aim for low cost.
- Food banks are used by people on very low incomes.

Availability of Food

- There is a vast array of foods to choose from including organic, multicultural and gluten-free.
- The best strategy for a healthy and varied diet is to plan a meal diary for the week, create a shopping list, then look for 'best buys'.

Seasonality

- Seasonal foods are foods that are harvested and consumed in the season they are naturally harvested in.
- In-season foods that are grown and sold locally should have more flavour and nutritional value than imported foods, e.g. English strawberries in June and July.

Enjoyment of Food

- Our enjoyment of food is affected by what the food looks, smells, tastes and feels like.
- We get maximum enjoyment from eating a variety of textures, flavours and colours.

Lifestyle

- Factors affecting household eating patterns include:
  - work (type of work done/hours worked)
  - travelling time
  - pastimes of individuals
  - who plans, prepares and cooks the meals.
- The amount of time available to prepare and cook food influences:
  - whether to use a microwave oven, a slow-cooker or a conventional oven.
  - the decision to use pre-prepared vegetables or from a greengrocer.
  - the type of food cooked, e.g. a casserole or a steak to fry.

Celebrations/Occasions

- Many religious festivals have strong eating traditions, e.g. hot cross buns at Easter, unleavened bread at Jewish Passover.
- Birthday parties or weddings generally include a selection of more expensive and extravagant foods as they are a special time for families.

Key Point

People need to make informed choices about the food they buy based on their income, lifestyle and preferences from the food available to them.

Quick Test

1. What does PAL mean?
2. What is the name of the government guide that advises the general public on how much to eat and what to eat?
3. To enjoy eating food, what should our meals contain a variety of?

Key Words

cardiovascular
Eatwell Guide
proportions
retailers

Review Questions

Microorganisms, Enzymes and Food Spoilage

1 Bacteria is a microorganism that causes food spoilage. Name **two** other microorganisms that cause food spoilage. [2]

2 State and explain **three** conditions needed for microorganisms to grow. [3]

3 Name **three** foods that spoil easily. [3]

4 Enzymes cause fruit and vegetables to ripen. How does ripening affect fruit? Tick (✓) **one** answer. [1]

a) They gradually lose their colour. ☐

b) They start to go brown. ☐

c) They may go green. ☐

d) They start to dry out ☐

Microorganisms in Food Production

1 Microorganisms are used in the manufacturing of certain foods. Explain how they are used in the production of bread. [3]

2 What is used in yogurt production to develop the taste and texture? [1]

3 Why have probiotic cultures become popular in dairy products? [3]

4 Explain why microorganisms are important in the production of blue-veined cheese. [2]

Bacterial Contamination

1 Bacteria in food can cause food poisoning. Name **two** high-risk foods. [2]

2 There has been an increase of food poisoning in recent years. Describe the main causes of food poisoning. [8]

Buying and Storing Food

1 Complete the following sentences about temperature controls. [4]

  a) Water boils at .... °C.

  b) The temperature of a freezer should be below .... °C.

  c) Bacteria multiply rapidly between .... °C and .... °C.

  d) When heating food, the internal temperature should reach .... °C.

2 Tick (✓) the correct box next to each statement to show if it is true or false. [3]

	True	False
a) Lettuce should be stored at the top of the refrigerator.		
b) Hot food should be placed in a refrigerator.		
c) Raw meat should be stored at the bottom of a refrigerator.		

3 Describe **four** ways of reducing the risk of food poisoning when storing food. [4]

....

....

....

....

....

....

....

....

....

Preparing and Cooking Food

1 Complete the following sentences using the correct temperature. [4]

| 5 °C | 63 °C | 75 °C | 18 °C |

a) Food should be stored in the refrigerator below ....

b) The core temperature of cooked food should reach ....

c) Cooked food should be kept out of the danger zone, which is between

.... and ....

2 What are the recommended temperatures for the following?

Reheating cooked foods: .... [1]

Storing chilled foods: .... [1]

3 Name an item of equipment used to check the temperature of foods when reheating. [1]

....

4 Describe the correct procedures to follow when using the equipment named in question 3. [6]

....

....

....

....

....

....

....

....

....

Practice Questions

Food Choices

1. Which of the following statements is **true**? Tick (✓) **one** answer. [1]

    a) Vegetarians include gelatine in their diet. ☐

    b) Kosher meat is eaten by Muslims. ☐

    c) Lacto-vegetarians eat honey. ☐

    d) Coeliac disease is caused by sensitivity to rice. ☐

2. <u>Underline</u> the correct answer. People with a severe allergy often carry their own medication in the form of… [1]

    an epi-pen.   insulin.   a ball-point pen.   Paracetamol.

3. State which nutrients may be missing from a vegan diet. [3]

4. Name **three** protein foods that **each** of the following types of vegetarian may include in their diet.

    Vegan: [3]

    Lacto: [3]

    Ovo-lacto: [3]

British and International Cuisines

1. Which of these cheeses is **not** a British cheese? Tick (✓) **one** answer. [1]

    a) Cheddar ☐   b) Brie ☐   c) Wensleydale ☐   d) Stilton ☐

2. Which of these Indian curry dishes was invented in the UK? Tick (✓) **one** answer. [1]

    a) Prawn Rogan Josh ☐   b) Chicken Tikka Massala ☐

    c) Lamb Khadery ☐   d) Chicken Dopiaza ☐

3. What **four** things make a British cheese distinctive? [4]

4 Add the components of a traditional British meal.

Protein: .... [1]

Starchy food: .... [2]

Vegetables: .... [3]

Sauce: .... [1]

Accompaniment: .... [1]

Sensory Evaluation

1 What is another way of describing 'umami' Tick (✓) **one** answer. [1]

a) A crunchy texture ☐ **b)** A savoury taste ☐

c) An unpleasant smell ☐ **d)** A sour taste ☐

2 Which sense is **least** likely to be needed when evaluating food? Tick (✓) **one** answer. [1]

a) Sight ☐ **b)** Smell ☐ **c)** Hearing ☐ **d)** Touch ☐

3 This is a chart to show the results of testing a pizza.

a) What name is given to this type of chart?

.... [1]

b) Identify the **best** feature of the pizza product tested. [1]

....

c) Identify the two **worst** features (areas to be improved). [2]

....

....

d) What **three** suggestions would you make to improve this product, and why? [6]

....

Food Labelling

1 Which of these items does **not** need to go onto a food label? Tick (✓) **one** answer. [1]

   a) Cooking instructions ☐   **b)** Nutritional information ☐

   c) Picture of product ☐   **d)** Date mark and storage ☐

2 What does 'e' stand for on a product label? Tick (✓) **one** answer. [1]

   a) European weight ☐   **b)** Estimated weight ☐

   c) Exceptional weight ☐   **d)** Extra volume ☐

3   **a)** Look at the chart below showing the traffic light labelling system.

Per 100g of food			
	Low	Medium	High
Fat	Less than 3g	3g – 20g	More than 20g
Saturated fat	Less than 1.5g	1.5g – 5g	More than 5g
Salt	Less than 0.3g	0.3g – 1.5g	More than 1.5g
Sugars	Less than 5g	5g – 15g	More than 15g

Adapted from the Food Standards Agency

Explain **six** ways in which the traffic light labelling system helps the consumer to make good food choices.

[6]

b) Name **three** things that must be included on pre-packed foods. [3]

_____

_____

_____

c) Explain **three** examples of when nutrition information must be provided on a pre-packed food product. [3]

_____

_____

_____

_____

Factors Affecting Food Choice

1 Which of the following food outlets would be used by people on a very low income? Tick (✓) **one** answer. [1]

a) Market stall ☐   **b)** Supermarket ☐

c) Food bank ☐   **d)** Grocer ☐

2 a) If you were responsible for feeding young children in your family, what eating model could you use to help you when planning meals? [1]

_____

b) Explain what needs to be considered when planning what to feed a family. Use the following headings and give **two** points for each heading.

What to cook: _____ [2]

_____

How to cook it: _____ [2]

_____

Time: _____ [2]

_____

Food and the Environment

You must be able to:

- Demonstrate knowledge and understanding of the environmental issues associated with food and its production.

Food Origins

- Seasonal foods are home-grown products that are traditionally grown or produced during particular seasons of the year, e.g. in the UK strawberries are harvested between June and September.
- **Transportation** development around the world has meant that when seasonal products are not available they can be imported from hotter climates where they are grown all year round.

Food Miles

- **Food miles** are the distance food travels from its point of origin to your table.
- The planes, boats and lorries used to transport food around the world all create carbon dioxide gas (CO_2), which is a contributory factor to global warming and **climate change** (**carbon footprint**).
- Food miles can be reduced by:
  - supporting British farmers and the economy – use farmers' markets, which showcase local and regional producers
  - eating seasonal products – our bodies get the right delivery of nutrients, minerals and trace elements that we need at the right time of year
  - being a wise shopper – purchase foods that have been produced nearer to Britain.

> **Key Point**
>
> Food and packaging waste contributes to greenhouse gases (GHGs).

Reducing Carbon Emissions

- **Recycling** and producing less waste also helps to reduce carbon emissions.
- Reducing the amount of **packaging** or using biodegradable packaging, which rots naturally, could help our environment.
- Recycling (re-using) using local collection facilities or bottle banks is 'green', however some of the chemicals used to clean recycled waste can cause pollution. Recycling may also use more energy than making packaging from new resources.

Food Waste and Landfill

- Nearly a third of all food produced ends up in **landfill** sites where it gives off methane gas as it decomposes. This gas adds to GHG emissions.

- **Food waste** can be reduced by:
  - using FIFO (First In First Out) storage
  - wise shopping and planning ahead
  - only preparing the food you need
  - using food before it goes out of date
  - using left-over food to make other dishes, e.g. mashed potato can be used to make croquettes for another day.
- Home **composting** is efficient, easy and clean. It benefits the garden and plant life by returning goodness to the soil – just food waste and the earthworms in the ground are needed.

Sustainable Food

- **Sustainable food** means food that will continue to be available for many years to come.
- Intensive farming can diminish the quality of food stocks for future generations.

Fruit and Vegetables

- Follow healthy eating guidelines and eat more fruit, vegetables, grains and pulses, and less animal protein.
- Home-grown garden or allotment fruit and vegetables provide a cost-effective variety of vegetables.

Fish

- Fish can be made more sustainable by:
  - restricting catch sizes
  - imposing minimum sizes of fish for sale
  - widening the selection of fish being eaten to more species
  - putting back young fish so that they can go on to breed and reproduce.
- Dolphin-friendly tuna makes the consumer aware that no dolphins have been accidently trapped in nets during fishing.
- Fishermen have allocated strips of ocean in order to fish sustainably in different areas of the world. There are set fishing regulations and quotas for their area.

Marine Stewardship Council – Seafood can be traced back to a certified sustainable fishery.

> **Key Point**
>
> Seasonal and sustainable foods address many environmental issues.

> **Key Words**
>
> transportation
> food miles
> climate change
> carbon footprint
> recycling
> packaging
> landfill
> food waste
> composting
> sustainable food

> **Quick Test**
>
> 1. Explain what food miles are.
> 2. How can recycling sometimes produce more pollution than making something from new?
> 3. Give **two** ways that fish stocks can be made more sustainable than intensive fishing.

Food Provenance and Production Methods

You must be able to:

- Demonstrate knowledge and understanding of where ingredients are grown, reared and caught
- Have a clear understanding of different farming methods and their effect on the environment.

Traceability

- **Traceability** means the ability to track any food, feed, food-producing animal or substance that will be used for consumption, through all stages of production, processing and distribution.
- This is so that when a risk is identified it can be traced back to its source in order to swiftly isolate the problem and prevent contaminated products from reaching consumers.

Modern Intensive Farming

- After the Second World War, farmers were offered subsidies to farm intensively to produce large scale, low cost products.
- This policy has resulted in:
  - fewer small farm communities
  - a greater number of larger business farms
  - large numbers of animals and poultry being kept in massive buildings and fed on high nutrient feeds in a short period of time, which is designed to maximise growth
  - the widespread use of antibiotics, growth enhancers, fertilisers and pesticides
  - small farm fields being opened up – woodland destroyed to make room for large machinery access.
- These methods are also employed all over the world, resulting in large surpluses of food being produced.

Manufacturer

Producer/grower

Distributor

Transportation

Retailer

Consumer

Responsibility for food safety

Farming Methods

Barn-reared Animals

- **Barn-reared animals** live in an environment similar to intensively-reared animals.
- They have access to natural light from windows.
- They live in a lower density of animals per square metre.
- They have access to environment enrichment such as fresh straw.

Organic Foods

- **Organic foods** are grown naturally without help from any chemical or synthetic treatments.
- They rely on natural compost and manure as fertilisers.
- Organic foods are not **Genetically Modified** – they are GM-free.

- There is no proof that organic food is more nutritious – buying organic food is a lifestyle choice.

Free-range Farming

- **Free-range farming** allows animals or poultry access to outdoor areas for part of their lives.
- Hens that are free-range produce eggs that are more nutritious and tasty. Animals reared this way also have better meat quality.
- Organic and free-range farming are more ethical and have a lower negative environmental impact.

Hydroponic Farming

- **Hydroponic** farming is the production of food using specially developed nutrient-rich liquids rather than soil.
- Hydroponic farming takes place in vast polytunnels or greenhouses in carefully controlled conditions.
- It is an expensive method, so is only used for high-value crops.

Fish Farming

- Increasing demand for fish has seen stocks diminishing in the wild through overfishing.
- The reduction of fish stocks may be due to lack of controls or the use of factory ships that strip the sea of every type of creature.
- **Hatcheries** release young fish into the wild.
- Some **fish farms** are on land, with the fish never exposed to natural resources.
- Fish farm tanks may be giant nets in fresh or seawater, where the fish are controlled, but still in a semi-natural environment.

Genetically Modified (GM) Foods

- This form of **intensive farming** is widely used in agriculture and the food processing industry. It is carefully controlled and regulated.
- GM foods are produced to be more resistant to plant disease, insect pesticides and viruses.
- The DNA in the product can be changed in order for the product to display particular characteristics, e.g. cattle with a higher milk yield, or sheep with a higher meat yield.
- As a result of higher yield, food products are cheaper and harvested in a shorter period of time.
- There are concerns about the use of GM in food production:
  - it is altering and playing with nature
  - possibilities for new strains of microorganisms to develop
  - there are potential risks to long term human (and animal) health and allergic reactions
  - pollen from GM and non-GM crops may potentially be mixed.

> ### Key Point
>
> Best quality protein foods are ones where the welfare of the animals has been considered.

> ### Key Words
>
> traceability
> field to fork
> barn-reared animals
> organic
> Genetically Modified (GM)
> free range farming
> hydroponics
> hatcheries
> fish farms
> intensive farming

> ### Quick Test
>
> 1. Why is it important that the origins of food can be traced?
> 2. What are the benefits of free-range farming?
> 3. Why is hydroponics an expensive farming method?

Sustainability of Food

You must be able to:

- Demonstrate knowledge and understanding of the impact that food has on local and global markets.

Carbon Emissions

- As our **Green House Gas (GHG)** emissions increase, the planet traps more energy from the Sun.
  - This damages the Earth's ozone layer.
  - This in turn causes changes to climate, and ultimately food and water supplies throughout the world.
- Livestock, especially cows, produce methane gas, which is 20 times more harmful than CO_2.

Global Climate Change Issues

- There are several issues arising from global climate change:
  - Air temperature and rainfall levels rising and falling have an effect on soil. Crops can easily fail as a result.
  - Flooding of areas of land, both coastal and inland.
  - CFCs have depleted the atmosphere's ozone layer and this can have a reduction in yield in some crops with ultraviolet radiation affecting them.
  - Photosynthesis of plants relies on carbon dioxide; increased cloud cover as a result of global warming restricts this from being as efficient.
  - Changes in climate can affect the pests that attack crops, and can change the foods that some helpful bugs use, e.g. bees dying and so not pollinating plants.

Tackling Sustainability of Food Sources

- There are ways that the sustainability of food sources can be addressed:
  - Prevent soil erosion from winds, high rainfall and flooding.
  - Look to improve crop varieties to match climate change, e.g. in drier climates grow crops that require less moisture.
  - Look at **crop rotation** to reduce soil erosion, and the general health of crops.
  - Put in irrigation systems in drier areas.
  - Increase crop diversity
  - Improve soil organics by using animal waste
  - Develop wind breaks.
  - Change the dependence on fossil fuels for transporting foods.

Key Point

Carbon emissions and global climate change affect food and water supplies.

- Tackle deforestation issues – trees remove CO_2 from the atmosphere. Large areas of forest are being cut down in order to graze animals or grow crops. As a result, CO_2 builds up, contributing to global warming, e.g. palm oil producers in Asia have cut down rainforest, which has also affected the habitat of many animals now in danger of extinction.

Sustainable Food

- Sustainable food is food that should be produced, processed, distributed and disposed of in ways that:
  - contribute to thriving local economies and sustainable livelihoods – both in the UK and, in the case of imported products, in producer countries
  - protect the diversity of both plants and animals and the welfare of farmed and wild species
  - avoid damaging or wasting natural resources or contributing to climate change.

Fairtrade

- **Fairtrade** is a foundation that pays a realistic income to farmers in developing countries.
- It ensures a fair price for the goods, giving a steady income, and covers the cost of sustainable production.
- It invests in the locality and in better working conditions.
- Many products now display the Fairtrade mark, e.g. bananas, tea, chocolate and coffee.

Food Assurance Schemes

- These are guaranteed standards of animal welfare or food safety that the consumer can rely upon.

Red Tractor

- **Red Tractor** is a logo that tells the consumer that the food has been produced, processed and packed to Red Tractor standards.
- The flag on the logo shows the country of origin of the food.
- Red Tractor assures…
  - standards of food hygiene and safety
  - standards of equipment used in production
  - animal health and welfare
  - environmental issues and responsible use of pesticides.
- Red Tractor is controlled and monitored by Assured Food Standards (AFS).
- Any product with this logo can be traced from farm to fork.

Food Production

You must be able to:

- Demonstrate a knowledge and understanding of the primary and secondary stages of food processing
- Demonstrate a knowledge and understanding of the production of two familiar food products.

Milk

- Milk comes from a variety of animals; in Britain we drink mainly cow's milk.
- Fresh milk has a layer of cream on top.
- **Homogenised** milk is forced through tiny holes in a machine. This breaks up the fat and disperses it, and it doesn't reform as a layer.
- Lactose intolerant people can substitute animal milk in their diet with milk made from soya, rice, coconut, almond or oat. These milks don't contain lactose.

Primary Processing

- Using **primary processing**, milk is processed to produce a variety of different types:
  - **Pasteurised** – this extends shelf life.
  - **Skimmed** – this is pasteurised but has had all or most of the fat removed.
  - **Semi-skimmed** – this is pasteurised but has had some of the fat removed.
  - **Ultra-Heat Treated (UHT)** – also known as 'long life', this has a shelf life of up to six months.
  - **Sterilised** – this has a longer shelf life, is homogenised and has a slightly caramel flavour.
  - Dried – does not need refrigeration until reconstituted; it's made by evaporating the water from the milk, which leaves a fine powder; it's non bulky to store.
  - Canned:
    - **evaporated** – milk that has had water evaporated off; it's sweet and concentrated, homogenised and is sealed in cans and sterilised
    - **condensed** – evaporated milk that hasn't been sterilised; it has added sugar and is very thick.

Secondary Processing

- **Secondary processing** can be used to process milk into other dairy products:
  - Cream – the fat removed from milk is used. Types of cream are single, whipping, double and extra thick. Cream can be further processed to make soured cream, clotted cream, and crème fraiche.
  - Butter – cream is churned to make butter. It can then be salted and made into regional varieties, e.g. ghee, continental.

Key Point

Primary processing is the initial process that the food product goes through for us to be able to use it.

Key Point

Secondary processing is a further process that can take place using the primary processed product, to make a new food product.

- Cheese – this is milk in its solid form. There are many regional and international varieties of cheese, depending on the methods or animal milks used to produce them.
- Yoghurt – milk has a bacteria culture added to it to make yoghurt. Probiotic products contain live bacteria that are beneficial to the digestive system.
- All dairy products need to be stored in the fridge, except dried and canned milks.

Wheat

- Wheat is the main cereal product. It is used in many countries around the world as it is quite easy to grow and is relatively cheap.
- Each grain of wheat is made up of different layers that have different functions and contain different nutrients.
- Different varieties are grown. Some are stronger than others.

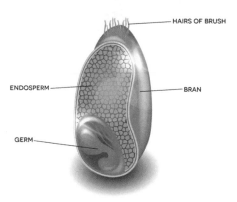

A grain of wheat

Primary Processing

- Wheat can be processed to produce a variety of flours.
- Wheat is made into a flour by a process called **milling**.
  - The grains are blended with other varieties and washed to remove grit and dirt.
  - Huge rotating rollers crush the grains at varying speeds.
  - The crushed grain is sieved and this produces flour.
  - If white flour is wanted, the bran is removed by further rolling.
  - The process is adjusted to produce the required type of flour.

Secondary Processing

- Flour can be further processed to make it into other products such as pasta or bread.
- **Pasta** is made from a variety of strong wheat called durum wheat.
- The flour is further processed by adding it to egg and making a paste, which can then be rolled and shaped to make a variety of pastas.
- **Bread** is made by mixing strong flour (which is high in gluten) with liquid and a raising agent such as yeast.

Quick Test

1. What is the outer skin of the wheat grain called?
2. What is homogenised milk?
3. Name **two** types of cream.
4. What type of flour is used to make pasta?

Food Processing

You must be able to:

- Know and understand how processing affects the sensory and nutritional properties of ingredients.

Preservation

- There are four main types of food **preservation**: high temperature; low temperature; **drying**; **chemical**.

High Temperature

Method	Details	Further Information
Canning	• Foods are placed in liquids in cans, sealed and heated to 121 °C • Long shelf life	• Loss of water-soluble vitamins C and B • Change in taste of the food
Irradiation	• Strictly controlled X-rays are passed through the food to delay ripening	• Vitamins A, C, E, and K may be lost • Food looks fresh and tastes the same
Pasteurisation	• Used mainly for milk, heating it to 71 °C for 15 seconds, then rapidly cooling it to 10 °C • Limited shelf life	• Little or no change to taste • Loss of vitamin B2 • Often fortified with vitamin D
Sterilisation	• Heated to 104 °C for 40 minutes or 115 °C for 15 minutes • Used mainly for milk and juices to prolong storage	• The process causes a slight caramelisation of the milk sugar content, resulting in a creamy flavour
Ultra-Heat Treatment (UHT)	• Heated to 140 °C for up to 5 seconds then put in an airtight container • Allows milk to be stored for up to six months	• There is a slight change in taste, colour remains similar and little change in nutrients

- Milk can also be preserved in cans – evaporated milk has water evaporated and is sterilised so it is much thicker and sweeter.
- Condensed milk is not sterilised and has added sugar so it is very sweet and thick.

Low Temperature

Method	Details	Further Information
Freezing	• Food is preserved for up to one year in temperatures between -18 °C and -29 °C	• Generally, no changes to food or nutrient content
Chilling	• This just extends shelf life	
Cold storage (CA)	• This just extends shelf life in an atmosphere of carbon dioxide	

Drying

Method	Details	Further Information
Sunlight	• An old method, which allows moisture to evaporate from the food in the Sun, e.g. fish, meats	• Dehydrating foods can affect colour, e.g. purple plums turn dark brown • They may develop a wrinkly surface/skin • The texture may change as well as the concentration of the flavour • Vitamins C and B6 (and others) may be lost
Oven drying	• Warm ovens are used to dry foods slowly	
Roller drying	• Used for foods that are reconstituted as 'instant' foods, e.g. baby foods	• Loss of vitamins A and C
Spray drying	• This method is used for some foods that are damaged by high heats	
Accelerated Freeze Drying (AFD)	• Food is frozen and the temperature is then increased to make the ice vaporise	• There is no change to the nutrient content of the food and flavour, colour and texture is mostly unaffected

Chemical

Method	Details	Further Information
Vinegar (pickling)	• The strong acid solution preserves the food, e.g. chutney, onions	• Changes to taste, flavour and texture • Shelf life is increased
Sugar	• Fruit is preserved, e.g. jam making	
Salt	• Meat and fish can be salted • Brine solutions can be used to preserve vegetables and canned fish	
Smoking	• Food is 'cooked' by exposing it to heat from wood fires	• Distinctive smoky taste
Alcohol	• Fruits are prepared and stored in brandy	• Changes to taste
Vacuum packaging	• Oxygen is removed from a sealed package, e.g. fish, cheese	• Long shelf life with fresh appearance and taste
Modified Atmospheric Packaging (MAP)	• Sealed packages have oxygen removed and the gas content inside changed, e.g. ready-prepared salads	

Quick Test

1. Which vitamins may be lost during irradiation?
2. What shelf life does UHT milk have?
3. How does vacuum packaging differ from MAP?

Key Words

preservation
temperature
drying
chemical preservation

Review Questions

Food Choices

1 Which religious dietary law forbids shellfish? Tick (✓) **one** answer. [1]

    a) Islam ☐   **b)** Hinduism ☐

    c) Sikhism ☐   **d)** Judaism ☐

2 What condition can be caused if a person has a serious allergy? [1]

3 What is lactose? [1]

4 **a)** Who is most likely to develop Type 2 diabetes? [2]

    b) Explain **two** main recommendations for a diabetic person's diet. [2]

5 What would be the effects of someone suffering from Coeliac disease eating a food containing gluten? [3]

6 What does insulin control in the body? [1]

British and International Cuisines

1 Which country would you associate sashimi with? Tick (✓) **one** answer. [1]

a) China ☐   **b)** Japan ☐   **c)** Spain ☐   **d)** China ☐

2 Explain **three** reasons why we now eat foods from other countries in the British diet. Use the following key words to help you.

Multicultural: _____ [1]

_____

Foods available: _____ [1]

_____

Lifestyle: _____ [1]

_____

Sensory Evaluation

1 Which of these statements describes a 'hedonic scale'? Tick (✓) **one** answer. [1]

a) How much someone likes something ☐   **c)** A row of faces ☐

b) How much someone dislikes something ☐   **d)** A list of words ☐

2 Name **three** sensory tasting characteristics. [3]

_____

_____

3 For each of the following, say which testing method would be the most suitable.

a) To find out which brand of fish fingers is preferred from a range of shops. [1]

_____

b) To find out whether people can tell the difference between low-fat cheddar cheese or full-fat cheddar. [1]

_____

4 Explain **two** ways the results of sensory testing be used to develop a food product. [2]

_____

_____

Food Labelling

1 **a)** How much does an average serving of this cereal weigh?

_____ [1]

b) How much energy does 100 g of this cereal provide?

_____ [1]

c) What could you serve with the cereal to make it healthier? [2]

Typical values	100g contains	45g serving contains
Energy	1570kJ 375kcal	710kJ 170kcal
Protein	10.3g	4.6g
Carbohydrate	73.8g	33.2g
of which sugars	15.0g	6.8g
Fat	2.0g	0.9g
of which saturates	0.3g	0.1g
Fibre‡‡	8.2g	3.7g
Sodium	0.2g	0.1g
Salt equivalent	0.6g	0.3g

‡‡Fibre has been determined by AOAC anal...
For guideline daily am...

_____

d) How much of the following nutrients would a serving provide?

Fibre: _____   Saturates: _____ [2]

2 Which **four** items on a food label are important for food safety? [4]

_____

_____

3 Which **two** date marks on fresh food tell the customer that the food is safe to eat?
Tick (✓) **two** answers. [2]

a) Sell-by date ☐   **b)** Use by date ☐

c) Display by date ☐   **d)** Best before date ☐

Factors Affecting Food Choice

1 What **four** factors affect our enjoyment of the food we eat? [4]

....

....

2 When feeding a family on a budget, which of the following foods is generally the most expensive to buy? Tick (✓) **one** answer. [1]

a) Chicken ☐ **b)** Potatoes ☐ **c)** Frozen peas ☐ **d)** Bread ☐

3 Explain **two** points about what influences the foods we choose to eat, using the following prompts:

Buying foods in season [2]

....

....

Choosing where to shop [2]

....

....

Cost of food [2]

....

....

Availability of different foods to buy [2]

....

....

4 Which of the following foods are associated with Easter? Tick (✓) **one** answer. [1]

a) Mince pies ☐ **b)** Hot cross buns ☐

c) Gingerbread ☐ **d)** Baklava ☐

Food and the Environment

1 Circle the correct options in the following sentences.

a) Planes, boats and lorries all create **carbon dioxide / carbon monoxide** emissions. [1]

b) Food miles are the distance food has travelled from its point of origin to **your supermarket / your table**. [1]

c) Heavily fertilised crops grown in the UK in heated greenhouses may be responsible for creating **less / more** carbon emissions than imported foods. [1]

d) FIFO places existing food stock to the **front / back / floor** of food storage areas. [1]

2 What is biodegradable packaging? [2]

3 Give **two** benefits of eating seasonal products. [2]

4 How can growing fruit and vegetables yourself help the environment? [3]

5 What is seasonality? [1]

Food Provenance and Production Methods

1 Which of these will help to make the food resource more sustainable? Tick (✓) **one** answer. [1]

a) Eat more animal proteins. ☐

b) Buy fish that are smaller. ☐

c) Eat a larger variety of fish species. ☐

d) Eat less fruit and vegetables. ☐

2 How would eating less beef and lamb help the environment? [2]

_____

_____

3 (Circle) the correct answer. [1]

Intensive farming produces food products that are **high cost / low cost**.

4 Name **three** things that are needed by organic farmers so that they can call their products 'organic'. [3]

_____

_____

Sustainability of Food

1 Explain crop rotation. [3]

_____

_____

2 What does GHG stand for? Tick (✓) **one** answer. [1]

a) Good Healthy Goods ☐ **b)** Global Heating Gas ☐

c) Green House Gas ☐ **d)** Genetically Healthier Groups ☐

3 Why is palm oil production damaging the environment? [5]

_____

_____

_____

4 Explain what a Fairtrade product is. [4]

_____

_____

_____

Food Production

1. What is the difference between fresh milk and homogenised milk? [2]

2. What does UHT stand for? [1]

3. Which type of milk is thicker? Circle the correct answer. [1]

   evaporated milk   **condensed milk**   **sterilised milk**

4. Explain what the term 'primary processing' means. [1]

5. Explain what the term 'secondary processing' means. [1]

Food Processing

1. Name **three** methods of preserving foods using high temperatures. [3]

2. a) Why are some foods irradiated? [1]

   b) Name **two** foods that could be irradiated. [2]

3　What does AFD stand for?　　　　　　　　　　　　　　　　　　　[1]

4　Name **one** benefit of low temperature storage.　　　　　　　　　[1]

5　In some hot countries it is traditional to dry foods to preserve them.

　a)　Explain **one** method of preserving food that is used in a hot country.　[1]

　b)　Give **one** example of the method you have chosen.　　　　　　[1]

　c)　Give details of **three** disadvantages of the method you have chosen.　[3]

6　a)　What does MAP stand for?　　　　　　　　　　　　　　　　[1]

　b)　Explain how MAP works.　　　　　　　　　　　　　　　　[1]

　c)　Give **two** examples of foods sold in this way.　　　　　　　[2]

　d)　Give **three** advantages of using this method.　　　　　　　[3]

Review Questions

Food and the Environment

1 Choose the correct words from the boxes and complete the following sentences. [6]

| food waste | green | recycling | energy | pollution | decomposes |

a) Some of the chemicals used to clean during .... can cause

....

b) Recycling may use more .... than making packaging from new resources.

c) .... can be reduced by wise shopping and planning ahead.

d) Food in landfill sites gives off methane gas as it ....

e) Recycling is a .... way of using collection facilities and bottle banks.

2 Explain how the fishing industry can ensure sustainability. [3]

....

....

....

....

3 What is a landfill? [1]

....

Food Provenance and Production Methods

1 Choose the correct words from the options given to complete the following sentences. [6]

| GM-free | barn-reared | organic | free-range | intensive farming | lifestyle |

a) .... produces large-scale low-cost food in sealed buildings, whereas

.... animals have access to natural light and some environmental enrichments.

b) .... grown foods are farmed naturally without any chemical or

synthetic treatments and are .... This is a .... choice.

.... farming allows animals to go outdoors for part of their lives.

2 What is 'traceability'? [5]

3 Is hydroponics an economical way of producing food? [1]

4 Name a more economical method of farming with less environmental impact. [1]

5 Give **one** disadvantage of the use of factory ships in a fishing fleet. [1]

Sustainability of Food

1 Which of these products are Fairtrade products? Tick (✓) the correct answers. [4]

a) Coffee ☐ b) Eggs ☐ c) Chocolate ☐

d) Flour ☐ e) Tea ☐ f) Bananas ☐

2 Products that have a Red Tractor logo on the packaging have a flag also to show the country of origin. [1]

True or **False**? ___

3 What is deforestation? [1]

4 Name **one** result of deforestation. [1]

Review Questions

Food Production

1. Is cheese a primary or secondary processed food? .... [1]

2. Name **two** milk products that don't need to be kept in the fridge. [2]

3. How is yoghurt made? [1]

4. Name **three** different types of cream. [3]

5. Name **one** use for:

   a) single cream. .... [1]

   b) clotted cream. .... [1]

   c) crème fraiche. .... [1]

6. Why is cheese difficult to digest? [1]

Food Processing

1. Name **two** vitamins that may be lost during the natural sunlight drying process. [2]

2. Which of the following are chemical preservation methods? Tick (✓) the correct answers. [2]

   a) Spray drying ☐   b) Salt ☐

   c) AFD ☐   d) Smoking ☐

3. Name **one** benefit of low-temperature storage. [1]

4 **a)** Name **three** methods of chemical preservation.   [3]

....

....

b) Explain how these methods of preservation are used.   [3]

....

....

....

....

c) What is the benefit of preserving foods?   [1]

....

5 You have been growing your own vegetables and have an excess of tomatoes and onions.
How can these be preserved to be used at a later date? Suggest two methods.   [2]

....

....

....

6 Name **two** methods that can be used to preserve ham or bacon.   [2]

....

....

Mixed Questions

1   The best method of cooking a quiche is: [1]

A   by steaming. ☐   B   by grilling. ☐

C   by microwave. ☐   D   by baking. ☐

2   Circle **one** answer. On a food product label, RI stands for: [1]

Requires Improvement

Reference Intake

Radical Impression

3   Give **three** functions of fat in the diet. [3]

4   When using raising agents, it is important to measure them correctly. Too little raising agent means a lack of rise and gives a close texture. What happens if too much raising agent is measured? [3]

5   When making a sauce it is important to understand heat transfer.

a)   Describe the method of heat transfer from the hob through the saucepan. [1]

b)   Describe the method of heat transfer through the milk. [1]

c)   Explain the purpose of stirring (agitating) the milk related to heat transfer. [2]

6 Name **two** groups of people who have an increased need for protein in their diets and explain why. [4]

....

....

....

7 What particular variety of wheat is used to make pasta flour? Why is this used? [1]

....

8 'Cuisine' relates to: [1]

   A the way in which food is cooked in a kitchen. ☐

   B the range of dishes and foods relating to a
      particular country or region. ☐

   C a brand name for kitchen equipment. ☐

9 Sterilised milk has a longer shelf life. Name **three** other types of milk that also have a longer shelf life and can be stored prior to use out of the fridge. Explain how each type of milk is packaged. [6]

....

....

....

....

10 **a)** Define the term 'coagulation' and describe how it affects the filling of a quiche. [3]

....

....

....

   b) Give another example of coagulation, apart from in a quiche filling. [1]

....

11 Some ingredients bring about specific characteristics when combined during cooking. Explain fully the function of the ingredient used in the following list of food products.

a) Wheat flour in a béchamel sauce. [3]

b) Rapeseed oil used to roast potatoes. [3]

c) Egg wash used to glaze pie pastry. [3]

d) Custard powder used to make vanilla custard. [3]

12 What are fish farms? Explain how they can help to make fish a more sustainable commodity. [2]

13 A trainee baker needs to recognise characteristics due to caramelisation.

a) She must tick the correct examples of caramelisation in the list. [2]

1 Sugars change colour and flavour. ☐

2 Starch turns golden brown. ☐

3 Sugar solution becomes syrup. ☐

4 Enzymic browning on bananas. ☐

b) She must also explain the problems that would occur if small sweet buns were cooked for
40 minutes at 210 °C (Gas 7). [3]

_____

_____

_____

14 Which of the following products would show a use-by date? Circle **one** answer. [1]

Packet of dried pasta

Can of soup

Pot of yoghurt

Jar of jam

15 At what temperature should pork be cooked to in order for it to be safe to eat? [1]

A 63 °C ☐ **B** 72 °C ☐

C 74 °C ☐ **D** 78 °C ☐

16 A student is investigating fat and has a list of dishes that contain fat. He needs to identify the
main function of fat in each dish. Complete the chart to show the main function of fat within
each dish. [5]

Name of Dish	Main Function of Fat in the Dish
a) Pastry pie crust	
b) Hollandaise sauce	
c) Glazed new potatoes	
d) Bread and butter pudding	
e) Steamed syrup sponge	

17 Which **one** of these fish is a white fish? [1]

A Prawn ☐ **B** Mackerel ☐

C Plaice ☐ **D** Scallop ☐

Mixed Questions

18 Explain intensive animal farming methods. [5]

_____

_____

_____

_____

_____

19 A yeast batter for rich dough needs to be fermented. Describe the conditions needed to enable fermentation to take place. [4]

_____

_____

_____

_____

20 Which of the following must be included on food labels by law? [3]

A Picture of the product ☐ **B** Allergen information ☐

C Lot or batch mark ☐ **D** Ingredients list ☐

21 **a)** Circle **one** answer. When looking at a traffic light food label, information is given for each food per: [1]

25 g **100 g** **200 g**

b) Circle **one** answer. The colours on the traffic light food label help the customer make healthier food choices. The healthiest food choice is shown as: [1]

Red **Amber** **Green**

22 What is the term used to describe holding a knife to cut an onion into two on a chopping board? [1]

A Hand ☐ **B** Bridge ☐ **C** Claw ☐ **D** Fist ☐

23 Obesity is considered a major risk to health. Underline **two** medical conditions that are linked to obesity. [2]

dementia   coronary heart disease   lung cancer

conjunctivitis   diabetes   dermatitis

24 When carrying out a tasting panel on a range of similar food products, which of these tests will decide which is the most popular product? [1]

A Rating test ☐   **B** Ranking test ☐

C Triangle test ☐   **D** Paired preference test ☐

25 Are these statements **true** or **false**? Write the answer next to each sentence. [3]

Chopitos consists of cubes of potato in a spicy tomato sauce. _____

Fish is common in traditional Japanese cuisine. _____

A traditional British meal consists of Chicken Tikka Massala with pilau rice. _____

26 Which of these proteins is **not** found in meat? Underline **one** answer. [1]

Elastin

Gluten

Myoglobin

Collagen

27 Identify the conditions that food spoilage organisms need to grow. [3]

_____

_____

_____

28 Which of the following points should be observed when setting up a tasting session? [2]

A Provide small samples of food. ☐

B Use clean spoons or forks for each sample tested. ☐

C Label foods clearly and accurately. ☐

D Give testers a piece of apple to eat between each sample. ☐

29 Which of the following products is an example of secondary processing? [1]

A Wholemeal flour ☐

B Bread ☐

C Potatoes ☐

D Eggs ☐

30 When filleting a flat fish:

a) What is the first step? .... [1]

b) What type of knife and chopping board would you need? [2]

....

c) Explain **three** points to show how you would know the fish is fresh. [3]

....

....

....

....

31 Foods that originate in an area of a country generally use locally sourced ingredients. Unravel the following **three** regional British foods. [3]

1 SINROCH TPYAS ....

2 BULDOE ROETSUGELC HSEEEC ....

3 SLECEC SKEAC ....

32 Fruit and vegetables are a good source of dietary fibre. Give **two** reasons why we need plenty of fibre in our diets. [2]

....

....

33 Which part of the body produces insulin? [1]

A Stomach ☐

B Liver ☐

C Pancreas ☐

D Spleen ☐

34 Blue veined cheese is one example of a food where microorganisms are used in food production. Name **one** other food that uses a microorganism in its production process. [1]

....

35 State **three** methods of preserving food at home. [3]

....

....

36 Tick the boxes below to show if the statements are **true** or **false**. [3]

Statement	True	False
Vegetarians eat fish		
Hindus eat pork		
Jewish people don't eat shellfish		

37 Fill in the missing word. [1]

Saturated fats raise the c.. level in the body.

38 Fillings and dental extractions in children have been linked to overconsumption of which type of carbohydrate? [1]

....

39 Which of the following statements describes the consistency of choux pastry dough? [1]

A Pouring ☐ B Heavy dropping ☐

C Coating ☐ D Soft ☐

40 a) Name **four** of the joints that can be found in a chicken. [4]

....

b) What colour of chopping board should be used to cut chicken? [1]

_____

c) What is the best choice of kitchen knife to use? _____ [1]

d) Explain the process of portioning out a chicken. [5]

_____

_____

_____

_____

_____

41 Which of the following nutrients is **not** found in fish? [1]

A Protein ☐

B Vitamin C ☐

C Zinc ☐

D Omega 3 fatty acid ☐

42 What advice would you give an elderly relative who is planning a healthy diet? [2]

_____

_____

43 Which of the following statements are **true**? [2]

Vegetarians may lack iron in their diet.	
Kosher meat is eaten by Muslims.	
Lacto vegetarians eat honey.	
Coeliac disease is caused by insensitivity to rice.	

44 Name **two** animals and the type of meat produced. [4]

Animal _____ Meat _____

Animal _____ Meat _____

45 Which of these dough mixtures does **not** use a strong flour? [1]

A Shortcrust pastry ☐   B Pasta ☐

C Choux ☐   D Bread ☐

46 Fruit and vegetables that are produced without the use of artificial fertilisers are called: [1]

A GM (genetically modified). ☐   B sustainable. ☐

C organic. ☐   D intensive farming. ☐

47 What is meant by the term 'food miles'? [1]

....

....

48 Explain the importance of using colour-coded equipment when handling and preparing food. [3]

....

....

....

49 What causes global warming? [1]

....

....

50 Diabetes has increased significantly in the UK over the past few years. [3]

Give **three** types of food you should avoid to decrease your risk of developing this disease.

....

....

51 Complete the chart below. Provide **three** examples of each type of protein. [6]

High Biological Value (HBV) Protein	Low Biological Value (LBV) Protein

52 Jake is preparing the following family meal: [8]

Chicken curry with rice

Chocolate mousse

Give Jake advice on how the high-risk ingredients in the meal should be prepared and cooked to avoid food poisoning.

53 Fill in the missing words. [3]

D_____ is controlled by careful management of s_____ in the diet, plus i_____ medication.

54 What do the following terms mean? [4]

Enriched bread dough

Strong elastic dough

Rolling boil

A dough

55 **a)** What is this logo called? [1]

....

....

b) Where would you be likely to find this logo? [1]

....

c) What does this logo tell you? [8]

....

....

....

....

....

....

....

....

56 Give **two** reasons why someone may need to increase their water intake. [2]

....

....

....

....

57 What is the name given to the production of foods using specially developed nutrient-rich liquids rather than soil? [1]

A Seasonal ☐   **B** Biodegradable ☐

C Hydroponics ☐   **D** Genetically Modified (GM) ☐

58 Give **two** functions of vitamin A in the body. [2]

59 Name **three** microorganisms that cause food to spoil. [3]

60 This is a recipe for bread.

200 g strong white flour

1 tablespoon vegetable oil

3 g yeast

3 g salt

150 ml warm water

How could the fibre content of this recipe be increased? [1]

61 Discuss how food poisoning can be prevented when preparing and cooking a chicken stir-fry. [6]

62 Which of these foods is least harmful to the environment? Circle the correct answer. [1]

Roast beef

Lamb stew

Beef burger

Roast chicken

63 Why is white bread a good source of calcium? [1]

....

64 Fill in the missing information. [4]

Method	Details on Main Function of Fat in the Dish
....	Used for foods which are reconstituted as 'instant' foods, e.g. baby foods.
Spray drying	This method is used for some foods that are ....
Accelerated Freeze Drying (AFD)	Food is .... and the temperature is then increased to make the ....

65 Airplanes, lorries and boats are all used to transport food. Which greenhouse gas do they produce? [1]

....

66 Explain what happens to food stored in a refrigerator section whose temperature is 10 °C. [6]

....

....

....

....

....

....

67 Vitamin C is an antioxidant. Which of the following statements best describes the function of an antioxidant? Underline the correct statement. [1]

Antioxidants protect us from pollutants in the environment.

Antioxidants protect us from food poisoning.

Antioxidants protect us from dementia in old age.

68 Look at the refrigerator below and label where the following foods should be stored. [4]

Eggs

Mango

Raw fish

Chocolate eclairs

69 Give **one** function of sodium in the diet. [1]

70 Which vitamin can be manufactured in the body by the action of sunlight on the skin? [1]

71 Vitamin C (ascorbic acid) is water soluble and easily lost in the cooking process. [1]

Which of the following is considered good practice to retain the vitamins in vegetable preparation?

A Prepare vegetables the day before you need them.

B Put the vegetables in cold water, then bring them to the boil.

C Cook vegetables very quickly in a small amount of boiling water.

D Make sure vegetables are very soft before draining and serving.

72 Explain what happens to food stored in a freezer operating at a temperature of −18 °C. [6]

73 Complete the following table. [4]

Vitamin	Main Function in the Body	Main Food Sources
Iron		
Calcium		

Total Marks _____ / 212

Answers

Page 7 Quick Test
1. With the point facing downwards
2. No
3. To avoid the food moving and the knife slipping
4. Bridge hold; claw grip

Page 9 Quick Test
1. Oily fish
2. **Any two from**: bright eyes, red gills; scales in place; layer of slime; firm flesh; smell of the sea
3. A blue chopping board and a filleting knife

Page 11 Quick Test
1. Covered, in the bottom of the fridge, so that the juices do not drip onto other foods
2. Long fibres
3. Animals; poultry; offal; game

Page 13 Quick Test
1. By hand
2. A mould or your hands
3. Sugar and water

Page 15 Quick Test
1. Strong flour
2. Carbon dioxide

Knife Skills
1. c) and d) [2]
2. **Any two from**: Jardinière [1] – baton shaped lengths [1]; Julienne [1] – matchstick-sized strips [1]; Macedoine [1] – medium-sized dice [1]; Chiffonade [1] – fine shreds of leafy vegetables [1]; Battonet [1] – Square-shaped lengths [1]; Turning [1]; – barrel shaped [1]
3. Julienne [1]; Battonet [1]; Jardinière [1]

Fish
1. c) [1]
2. d) [1]
3. a) **Any one of**: Cod [1]; Haddock [1]; Whiting [1]; Pollock [1]; Coley [1].
   b) **Any one of**: Plaice [1]; Sole [1]; Halibut [1]; Turbot [1].
   c) **Any one of**: Salmon [1]; Sardines [1]; Herring [1]; Mackerel [1]; Tuna [1]; Whitebait [1].
4. **Possible responses:** Check that the eyes are bright – dull eyes indicate that the fish may still be safe to eat but is not very fresh [1]; Check that the scales are still in place [1]; Check that the gills are bright red – if they are dark red, the fish is not fresh [1]; Smell the fish – does it have a slightly salty smell of the sea? This is an indicator of freshness, a nasty smell will not improve with cooking [1]; Check for a thin layer of clear slime – if the liquid is not clear, the fish has begun to rot [1]; Check the flesh is firm with your finger – if the flesh bounces back, it is fresh [1]. **(1 mark for each point made, up to a maximum of 5 marks.)**

Meat
1. b) [1]
2. d) [1]
3. a) **Any two from**: Stewing [1]; Braising [1]; Pot roasting [1]; Casseroling [1].
   b) Grilling [1]; Frying [1]
   c) Grilling [1]; Frying [1]
4.

Temperature check	Time per 500 g
75–80 °C	**30** mins [1] + 15 mins
75–80 °C	20 mins + **20** mins [1]
Rare: 52 °C Medium: **60 °C** [1] **Well done** [1]: 75–80 °C	20 mins 25 mins 30 mins

5. **Any two from**: Pork [1]; Chicken [1]; Offal [1]. **(Meat products, e.g. sausages, burgers, kebabs, not allowed.)**

Prepare, Combine and Shape
1. c) [1]
2.

Method	Food
Grating	Cheese [1]; Carrot [1]
Chopping	Herbs [1]; Onion [1]
Slicing	**Any two from**: Carrot [1]; Onion [1]; Bread [1]
De-coring	Apple [1]; Pineapple [1]

Dough
1. b) [1]
2. b) [1]
3. Durum wheat [1]
4. Glutenin [1]; Gliadin [1]
5. **Possible responses**: Spinach [1]; Beetroot [1]; Tomato [1]; Squid ink [1]; any other suitable answer [1]. **(1 mark for each food, up to a maximum of 2 marks.)**
6. Rest it [1]; After kneading [1]
7. a) Weigh all the ingredients accurately. [1]
   b) Add flour immediately to the pan. [1]
   c) Add eggs and beat in gradually. [1]
   d) Consistency must achieve heavy dropping. [1]
   e) Bake in a hot oven. [1]

Page 21 Quick Test
1. Source of vitamins; protection for organs; A component of hormones; Energy
2. Amino acids
3. Beans on toast/hummus and pitta bread

Page 23 Quick Test
1. Carbohydrate provides energy
2. Excess carbohydrate is converted to fat and is stored under the skin

3. Non-Starch Polysaccharides (this is dietary fibre)

Page 25 Quick Test
1. Vitamins A, D, E and K
2. **Any three from**: Citrus fruit; Kiwi fruit; Blackcurrants; Salad and green vegetables; Potatoes
3. Folic acid/folate

Page 27 Quick Test
1. To replace supplies lost through blood during periods
2. When bones are at maximum strength
3. Hormones in the thyroid gland
4. **Any two from**: Milk and dairy foods; Green leafy vegetables; Nuts, seeds and lentils; White bread – calcium is added by law; Wholegrain cereals

Page 29 Quick Test
1. **Any three from**: Base your meals on starchy carbohydrates; Eat lots of fruit and vegetables (5–7 portions per day); Eat plenty of fish, including oily fish; Cut down on saturated fat and sugars; Eat less salt – no more than 6g a day for adults; Get active and be a healthy weight; Don't get thirsty (drink six to eight glasses of water a day); Don't skip breakfast
2. For the development of the neural tube of the foetus
3. The first breast milk, which is full of antibodies
4. Weaning
5. A condition where bones become weak and break easily

Page 31 Quick Test
1. Bones that become brittle and break easily
2. Lack of energy; Pale complexion; Shortness of breath; Tiredness.

Knife Skills
1. Name of knife, **any three from**: Cook's [1]; Paring [1]; Boning [1]; Filleting [1]; Carving [1]; Bread [1]; Palette [1].
   Description: Cook's: Come in different sizes and their strong, ridged blades make them suitable for a wide range of tasks [1]; Paring: A small knife with a thin and slightly flexible blade [1]; Boning: A very strong blade that will not bend or break easily, can be straight or curved [1]; Filleting: Thin-bladed, flexible and very sharp knife [1]; Carving: Long blades with a serrated or plain edge, can be rounded or pointed [1]; Bread: Long serrated edge [1]; Palette: Flexible blade, rounded at the top [1].
   One possible use: Cook's, **any one of**: Dicing [1]; Chopping and trimming vegetables [1]; Chopping and trimming meat and poultry [1]; Chopping fresh herbs [1]. Paring: Fruit and vegetable preparation [1]. Boning: Removing bones from meat joints and poultry [1]. Filleting: Filleting fish [1]. Carving: Carving meat joints or cooked hams [1]. Bread: Slicing loaves and other

bread products, sandwiches [1]. Palette, **any one of**: Icing cakes [1]; Turning food over during cooking [1]; Moulding and smoothing food [1].

Fish
1. d) [1]
2. b) [1]
3. Method, **any one of**: Canning [1]; Freezing [1]; Smoking [1]; Salting [1]. **(1 mark for each method, up to a maximum of 3 marks.)**
   Canning example, **any one of**: Salmon [1]; Tuna [1]; Sardine [1]; Pilchard [1].
   Freezing example, **any one of**: Cod [1]; Haddock [1]; Pollock [1]; Coley [1].
   Smoking example, **any one of**: Salmon [1]; Haddock [1]; Mackerel [1].
   Salting example: Cod [1]; Herring [1].
   (1 mark for each example of method selected, up to a maximum of 3 marks.)
4. Smoking [1]; Salting [1]
5. 1 Protein [1]; 2 Minerals, **any one of**: iron [1]; zinc [1]; iodine [1]; 3 Vitamins, **any one of**: A [1]; D [1]; 4 Omega 3 fatty acids [1].

Meat
1. d) [1]
2. b) [1]
3. a) Pork [1]
   b) Beef [1]
   c) Lamb/mutton [1]
   d) Venison/game meat [1]
   e) **Any one of**: Chicken [1]; Turkey [1]; Duck [1]; Goose [1].
4. **Any three from**: Protein [1]; Fat (saturated) [1]; Iron [1]; Calcium [1]; Phosphorous [1].
5. **Any three from**: Raw meat should be kept separate from cooked meat to avoid cross-contamination [1]; Meat should be stored covered at the bottom of the fridge so that if meat juices leak out they cannot drip onto other foods [1]; Chilled meat should be stored at between 0 °C and 5 °C [1]; Frozen meat should be stored at between –18 °C and –22 °C [1].

Prepare, Combine and Shape
1. a) [1]
2. b) [1]
3. b) [1]
4. A milk glaze (egg glaze also acceptable) [1]
5. **Any two from**: Loaf [1]; Cake [1]; Jelly [1]; Burger [1]; Mousse [1].
6. **Any two from**: Consistency in depth [1]; Consistency in size [1]; Consistency in shape [1].

Dough
1. d) [1]
2. a) 1:2 [1]
   b) Plain flour [1]
   c) Mixture of lard [1]; vegetable fat/butter [1]
   d) Rubbing-in [1]
   e) **Any one of**: short/crisp/light [1]
3. a) Acts as a raising agent [1]
   b) Feeds the yeast [1]
   c) Increases shelf-life/can add colour and flavour [1]
   d) Provides the yeast with moisture/helps to form gluten [1]
4. It allows the gluten [1] in dough to relax [1] before it is rolled out.
5. c) [1]

Protein and Fat
1. c) [1]
2. b) [1]
3. Possible answers, **any one of**: Beans on toast [1]; Lentil soup with bread [1]; Hummus and pitta bread [1]; or any other suitable answer [1].
4. b) [1]
5. **Any two from**: Full fat milk [1]; Hard margarine [1]; Coconut oil [1]; Palm oil [1]; Butter [1]; Egg [1]; Cheese [1]; Cream [1]; Cakes [1]; Biscuits [1]; Chocolate [1]; Pastries [1]; Meat products [1]; Beef fat/dripping/lard [1]; or any other suitable answer [1].

Carbohydrate
1. c) [1]
2. **Any three from**: Bread [1]; Potatoes [1]; Pasta [1]; Rice [1]; Breakfast cereals [1]; or any other suitable answer [1].
3. **Any two from**: Carbohydrates are needed by the body because they are the main source of energy in the body for movement [1]; Carbohydrates are needed by the body for digestion [1]; Carbohydrates are needed by the body for growth [1].
4. Non Starch Polysaccharide [1]
5. **Any three from**: Wholegrain cereals [1]; Wholemeal bread [1]; Wholegrain breakfast cereals, e.g. bran flakes, weetabix, shredded wheat, porridge oats [1]; Wholemeal pasta [1]; Wholemeal flour [1]; Fruit [1]; Vegetables [1]; Dried fruit [1]; Nuts [1]; Seeds [1]; Beans [1]; Peas [1]; Lentils [1]; or any other suitable answer [1].

Vitamins
1. Vitamin B group [1]; Vitamin C [1]
2. c) [1]
3. **Any four from**: Don't prepare the vegetables too early – the longer fully cooked vegetables sit around before being eaten, the more nutrients are lost [1]; Use boiling water when starting off – the vegetables will cook more quickly and more of their nutrients will be retained [1]; Use a minimal amount of water to boil vegetables so that any nutrients that leach out are concentrated in a smaller quantity of liquid, that can then be made into a sauce or gravy [1]; Do not overcook/leave vegetables al dente, then nutrients won't leach out in the first place [1]; Use the cooking liquid to make a sauce/gravy so that any nutrients that have leached into the water during cooking are retained [1]; Use a sharp knife and do not over cut – the larger a surface area that can be retained, the more goodness will be locked in [1].
4. Fat-soluble vitamins [1]
5. From the action of sunlight on the skin [1]; Which enables the body to make vitamin D [1].

Minerals and Water
1. b) [1]
2. **Any two from**: Cheese or a named cheese [1]; Cream [1]; Yoghurt [1]; Bread [1]; Fish or fish where the bones are eaten [1]; Leafy green vegetables or named vegetable [1]; Nuts or named nut [1]; Seeds or named

seed [1]; Dried fruit [1]; or any other suitable answer [1].
3. Iron supports the production of haemoglobin in red blood cells [1]; Which transports oxygen around the body [1].
4. **Any two from**: Red meat – liver [1]; Lentils [1]; Dried apricots [1]; Cocoa [1]; Chocolate [1]; Corned beef [1]; Curry spices [1]; Green leafy vegetables [1]; Fortified breakfast cereals [1]; or any other suitable answer [1].
5. **Any three from**: Water is essential for normal brain function [1]; Drinking enough water decreases the risk of kidney problems [1]; Drinking the right amount of water helps to maintain normal blood pressure [1]; Water aids bowel movements [1]; Water helps to maintain cell function/healthy skin [1]; The right amount of water in the body regulates temperature [1]; Water is needed to maintain hydration [1]; Water aids digestion [1]; Water is needed to make body fluids, e.g. blood, saliva and mucus membranes [1].

Making Informed Choices
1. Eatwell Guide [1]
2. **Any three from**: Eat less sugar [1]; Eat more fibre [1]; Eat more starchy food [1]; Eat at least five portions of fruit and vegetables daily [1]; Don't skip breakfast [1]; Eat the right amount to stay a healthy weight [1]; Follow the Eatwell Guide [1]; Drink plenty of water [1].
3. Weaning [1]
4. Carbohydrates [1]

Diet, Nutrition and Health
1. b) [1]
2. b) [1]
3. **Any two from**: Coronary heart disease [1]; Type 2 Diabetes [1]; Some types of cancer [1]; High blood pressure [1]; Risk of stroke [1]; Arthritis [1]; Breathing problems [1].
4. Insulin [1]
5. **Any three from**: Weight gain/obesity [1]; Can produce high/bad cholesterol [1]; Can block arteries [1]; Angina [1]; Coronary Heart Disease (CHD)/heart disease/heart attack [1]; Trans fats/higher risk of cancer [1].
6. BMR is the body's basic energy rate if the person is at complete rest [1]; BMR is the energy needed by the body just to function, with no movement except breathing/natural functions at rest [1].

Page 41 Quick Test
1. True
2. Convection
3. Water-based; Fat-based; Dry
4. Conduction; Convection; Radiation

Page 43 Quick Test
1. Enzymes
2. Coagulation/Denaturation

Page 45 Quick Test
1. Gelatinisation
2. Radiation/grill
3. True

Page 47 Quick Test
1. Shortening
2. True
3. Creaming

Page 49 Quick Test
1. Air is the raising agent
2. Air expands when heated
3. By forming a foam
4. By turning to steam

Pages 50-53 **Review Questions**

Protein and Fat
1. chicken [1]; milk [1]; soya [1]
2. Protein complementation [1]
3. The level of cholesterol [1]

Carbohydrate
1. pasta [1]; potato [1]; oats [1]
2. lentils [1]; cabbage [1]; apples [1]; shredded wheat [1]
3. Sugar [1]

Vitamins
1. Kiwi fruit [1]; lemons [1]; cabbage [1]
2. Folate (folic acid) reduces the risk of neural tube defect (Spina Bifida) in babies, so pregnant women should take a folate (folic acid) supplement [1].
3. **Any two from**: Vitamin C and vitamin B are destroyed by heat so vegetables that are not subjected to heat retain these vitamins [1]; Vitamin C is destroyed on exposure to oxygen therefore after it has been cut, before cooking, there are losses [1]; Vitamin C and vitamin B are water-soluble so will leach into water during the cooking process [1]
4. Ascorbic acid [1]

Minerals and Water
1. haemoglobin [1]; red blood [1]; oxygen [1]
2. Due to monthly menstrual blood loss [1]
3. The body's metabolic rate [1]
4. **Any two from**: In hot weather to replace water lost by excess sweating [1]; After physical exertion to replace water lost by excess sweating [1]; Lactating mothers need an increased supply of water for the production of milk [1]; People suffering from sickness and diarrhoea need to increase their intake of water in order to keep hydrated due to loss of water via the sickness and diarrhoea [1].

Making Informed Choices
1. chocolate [1]; beef [1]; apricots [1]
2. c) [1]
3. iron [1]; liver [1]

Diet, Nutrition and Health
1. a) [1]
2. milky drinks [1]; cheese [1]; yoghurts [1]
3. Diabetes is controlled by careful management of sugar in the diet [1]; In addition, insulin medication may be frequently injected [1].
4. c) [1]
5. In old age bones become weak and brittle and are more likely to break – this is when the condition of osteoporosis is most likely to occur [1].

Pages 54-57 **Practice Questions**

Cooking of Food, Heat Transfer and Selecting Appropriate Cooking Methods
1. a) microorganisms [1]
   b) shelf life [1]; keep quality [1]
   c) variety [1]
   d) digest [1]
2. Conduction [1]; Convection [1]; Radiation [1]
3. Water-based methods, **any two from**: Boiling [1]; Stewing [1]; Steaming [1]; Poaching [1]; Braising [1]; Pressure cooking [1].
   Dry cooking methods, **any two from**: Baking [1]; Grilling, [1]; Roasting [1]; Barbecuing [1]; Dry-frying [1].
   Fat-based **any two from**: Shallow frying in fat [1]; Deep-fat frying [1]; Stir frying, etc. [1].

Proteins and Enzymic Browning
1. Heat [1]; Reduction of pH [1]; Enzymes [1]; Mechanical actions [1]
2. a) It changes from being runny to being set [1]
   b) Due to coagulation [1]; Due to the effect of heat [1]
3. a) Gluten [1]
   b) Strong flour contains more protein than soft cake flour [1].
4. a) Water must be added [1]; Kneading must take place [1]
   b) Gliadin [1]
5. a) Enzymic browning [1]
   b) **Any one of**: Inactivate enzymes by using an acid such as lemon juice [1]; Remove the presence of air, e.g place in water [1]; Cook [1].

Carbohydrates
1. liquid [1]; gelatinisation [1]; heated [1]; stirred [1]; boiling point [1]
2. a) Colour: The bread turns golden [1]
   Flavour: The bread tastes sweeter [1]
   Texture: The bread becomes crisp [1]
   b) Dextrinisation [1]
   c) Hygroscopic [1]
   d) The toast becomes less crisp [1]

Fats and Oils
1. a) [1]
2. a) Functions: Aerate mixture [1]; Trap air [1]; Characteristics: Cake rises [1]; Increased volume [1]
   b) Function: Shorten the dough [1]; Characteristic: Crumbly short texture [1]

Raising Agents
1. To increase volume [1]; To make a light texture [1]; To raise mixtures [1]
2. a) Air [1]
   b) The raising agent is incorporated by whisking air into the sponge mixture, either by hand or with an electric mixer [1]
   c) **Any one of**: Baking powder [1]; Bicarbonate of soda [1]; Cream of tartar [1].
   d) Carbon dioxide [1]
   e) During baking the heat of the oven causes the raising agents to be active and produce the gas [1]; The gas swells with heat and raises the scone [1].

Pages 58-67 **Revise Questions**

Page 59 Quick Test
1. Bacteria
2. Between 5 °C and 63 °C – this is known as the 'danger zone'
3. Eat food as soon as possible after making/cooking, if not cool down as quickly as possible and store in the fridge or freezer
4. Possible answers: Bread; Wine; Beer

Page 61 Quick Test
1. Lactobacillus
2. Mould
3. Safeguard the immune system; Improve digestion; Keep the body's intestine balanced

Page 63 Quick Test
1. Any or all of the following are possible: Stomach pains; Diarrhoea; Vomiting; Nausea (feeling sick); Fever
2. **Any three from**: Salmonella; Staphylococcus Aureus; Clostridium perfingens; Clostridium Botulinum; Bacillus cereus; E-Coli; Listeria; Campylobacter; Norovirus
3. Raw food; Bacteria passed on from people; Bacteria carried in air and dust; Bacteria on equipment and utensils; Soil; Pests; Contaminated water; Degrading food waste
4. Very young children and babies; Elderly people; People who have a serious illness or are recovering from a serious illness; Pregnant women and nursing mothers; People who have allergies

Page 65 Quick Test
1. The 'Use by' and 'Best before' dates
2. Between 1 °C and 4 °C to make sure that food is held below 5 °C
3. False
4. −18 °C or below
5. Ice crystals remain at the centre; Cooking will melt the ice but the correct core temperature may not be achieved and bacteria may survive

Page 67 Quick Test
1. **Any four from**: Before preparing food; Between handling raw and high-risk or ready-to-eat foods; After you have been to the toilet; After sneezing or coughing; After changing a waterproof plaster; After using cleaning fluids; After handling food waste or rubbish; After handling known allergens
2. The thickest part of the food (core) should reach 75 °C, or 70 °C for two minutes

Pages 68-72 **Review Questions**

Cooking of Food, Heat Transfer and Selecting Appropriate Cooking Methods
1. In the following order: flavours [1]; mouthfeel/texture [1]; texture/mouthfeel [1]; bulk [1]; colour [1].
2. Conduction occurs as heat transfers through the pan [1]; This heat comes from the heat source [1]; Convection takes place in water in a pan during boiling [1]; Heat passes through the potato by conduction [1]; Boiling uses two methods of heat

transfer, the first method is conduction [1]; The second method is convection [1].

Proteins and Enzymic Browning
1. Meaning: Protein denaturation means the functional properties [1]; of a protein change [1].
   Example, possible answer: An example is cooked egg white [1]; It is denatured and therefore it will not whip to a foam [1].
2. In the following order: gluten [1]; elastic [1]; kneading [1]; short [1]; structure [1]

Carbohydrates
1. The solution becomes viscous when water and sugar are heated together [1]; The colour of this liquid becomes increasingly golden [1]
2. Flour: Dextrinisation   [1]
   Sugar: Caramelisation   [1]
3. Dextrinisation, possible answers, **any one of**: Brulee [1]; Crème caramel [1]; Spun sugar [1]; Sweet making [1]; Sticky ribs or wings [1] or any other suitable answer [1]. Caramelisation, possible answers, **any one of**: Toast [1]; Au gratin [1] or any other suitable answer [1].
4. a) Choice: Recipe B   [1]
   b) Explanation: The amount of flour would thicken the milk [1]; To make a smooth coating sauce [1]; The amount of flour in Recipe A would not [1]; be enough, the sauce would remain runny [1].

Fats and Oils
1. a) The rubbing-in method   [1]
   b) Fat coats the flour grains – the fingertips are used [1]; Added water does not penetrate the flour grains [1]; Therefore, gluten [1] development is limited [1]; Pastry not kneaded [1]. Cooking denatures the gluten and the pastry texture is crumbly, crisp and short [1].
2. In the following order: Emulsions [1]; immiscible [1]; separates [1]; stable [1]; lecithin [1]; natural [1]
3. a) True
   b) False
   c) True
   d) True
4. a) The margarine/butter [1]; and the caster/baking sugar [1]
   b) The mixture becomes paler as the fat and sugar are creamed together [1]
   c) Trapping air (aeration) [1]; Causes a paler coloured air-in-fat foam to form [1]
   d) During baking the trapped air would expand [1]; This makes the cakes rise [1]

Raising Agents
1. 1 A warm temperature is needed [1]; Of 25 °C–35 °C [1]
   2 Moisture is needed [1]; This is supplied from the liquid in the recipe [1]
   3 The yeast requires food [1]; This is supplied by the flour or the added sugar [1]
   4 Time is needed [1]; To produce the gas that raises the mixture [1]
2. a) Raising agents are active ingredients that make food products rise or processes that make food products rise [1]; and have open airy textures

[1]; Raising agents can be biological, chemical or physical [1]
   b) Name of product, possible answers, **any three from**: Scones [1]; Yeasted products [1]; Choux pastry [1]; Batters [1]; Whisked sponge [1] or any other suitable answer [1]. **[The correct raising agent must then be matched to each product to gain the mark.]**
   Name of raising agent, **any three from**: Scones: baking powder/bicarbonate of soda [1]; Yeasted products: yeast [1]; Choux pastry: high moisture content provides steam during the baking process that raises the mixture [1]; Batters: air whipped in during mixing process/baking powder/bicarbonate of sode/cream of tartar [1]; Whisked sponge: air whisked in during mixing process [1].
3. a) A sieve can be used to pass air through the flour [1].
   b) Whisking, by hand or with and electric mixer, distributes air through the mixture [1]
   c) Both of these methods add air [1]

Microorganisms, Enzymes and Food Spoilage
1. Enzymes are molecules that control chemical reactions in food [1]
2. Enzymic browning   [1]
3. Apple/banana   [1]
4. **Any one of**: Coat the fruit in an acid such as lemon/lime juice [1]; Coat the fruit in sugar to halt/prevent browning [1]; Stop oxygen from turning the fruit brown by immersing the fruit in water or juice [1]

Microorganisms in Food Production
1. b)   [1]
2. Lactic acid   [1]
3. False   [1]

Bacterial Contamination
1. True   [1]
2. a)   [1]
3. It means that the food is safe to eat up until the date marked on the food product or its packaging [1]

Buying and Storing Food
1. True   [1]
2. a) cooked   [1]
   b) perishable   [1]
   c) 0 °C and 4 °C, below 5 °C   [1]
   d) −18 °C and below   [1]

Preparing and Cooking Food
1. c)   [1]
2. If food is not reheated properly, the bacteria in it may survive and multiply – every time food is reheated the chances of this happening are increased [1]
3. cooked rice [1]; cooked chicken [1]

Page 77 Quick Test
1. Judaism and Islam
2. A vegan does not eat honey

3. Foods that contain gluten, e.g. wheat, barley, rye, oats

Page 79 Quick Test
1. Possible answers: Cornish Blue (from Cornwall), Lanark Blue (from Scotland), Caerphilly (from Wales)
2. Roast meat, beans, pulses and fish all provide protein. Roast potatoes (starchy carbohydrate) and accompanying vegetables provide at least two of the five recommended portions of vegetables per day.
3. The crimped edge of a Cornish pasty provided a 'baked-in' handle for tin miners to hold while they ate the pasty. Sometimes they threw the 'handle' away and sometimes they ate it.

Page 81 Quick Test
1. Texture
2. On your tongue
3. So that the tester is not influenced by brand loyalty

Page 83 Quick Test
1. 100 g
2. European Union
3. The following must also be included: Kilocalories and kilojoules; Protein; Sugars

Page 85 Quick Test
1. PAL stands for Physical Activity Level
2. The Eatwell Guide
3. A variety of textures, flavours and colours

Microprganisms, Enzymes and Food Spoilage
1. **Any two from**: Yeast [1]; Mould [1]; Fungi/fungus [1].
2. **Any three from**: The length of time exposed – the longer the exposure, the greater the growth [1]; Temperature – the higher the temperature, the faster and greater the growth [1]; Moisture – the higher the level of moisture the greater the growth [1]; The amount of microorganisms originating in the food itself influence growth – the more microorganisms there are to begin with, the more rapid their growth [1]; Level of oxygen/air/air flow – the more exposure to oxygen, the greater the chance of growth [1]; Neutral pH [1].
3. **Any three from**: Milk/yogurt [1]; Meat/poultry [1]; Fish/shellfish [1]; Eggs [1]; Vegetables [1]; Salad [1]; Fruit [1]; Bread [1]; Cream [1]; or any other suitable answer [1].
4. b)   [1]

Microorganisms in Food Production
1. **Any three from**: Bread dough is mixed and kneaded, millions of air bubbles are trapped and distributed through the dough [1]; The yeast in the dough provides microorganisms, which absorb the starches and sugars in the flour, turning them into alcohol and carbon dioxide gas [1]; This gas expands the air bubbles, causing the bread to rise [1]; During rising, the yeast divides and multiplies, producing more carbon dioxide. As long as there is ample air and

food (carbohydrates) in the dough, the yeast will multiply until its activity is stopped by the heat of the oven **[1]**.
2. A culture is used to develop taste and texture **[1]**.
3. Probiotic cultures offer health benefits such as helping to improve digestion **[1]**; Safeguarding the immune system **[1]**; Keeping the body's intestinal flora in balance **[1]**.
4. The cheese is treated with mould, which contains microorganisms, and these microorganisms develop as the cheese matures **[1]**; The mould continues to develop throughout the process, imparting the blue-veined appearance of the cheese and giving flavour **[1]**.

Bacterial Contamination
1. **Any two from**: Poultry – chicken/duck/turkey/goose **[1]**; Meat – beef/lamb/pork/sausages/burger **[1]**; Offal – liver/kidneys **[1]**; Fish/shellfish **[1]**; Eggs **[1]**; Dairy Products – cream/milk/yogurt **[1]**; Cooked rice **(no marks awarded for 'rice' only)**/reheated rice **[1]**; Custards/sauces/gravy **[1]**; Credit will also be given for a named food product, e.g. chicken curry containing fresh cream **[1]**.
2. **Any eight from**: The presence of excess microorganisms in food are a major cause of food poisoning – usually bacteria but sometimes viruses **[1]**; Food that has not been cooked properly, e.g. eggs that are still raw/undercooked products, i.e. meats **[1]**; Food that has not been stored properly, e.g. where cross contamination has taken place between raw and cooked meats **[1]**; Food poisoning takes place when microorganisms are allowed to multiply at a high rate, this is when they produce toxins/poisons/waste products (microorganisms cause the symptoms of food poisoning) **[1]**; A lack of knowledge and understanding about how to store, cook and prepare foods in the home and or in industry can lead to food poisoning **[1]**; Intensive farming methods can lead to disease spreading easily in perishable foods such as poultry or seafood, thus leading to food poisioning **[1]**; Dirty water can be a cause of food poisoning as the water can carry bacteria into the body **[1]**; Pests/animals can transmit food posioning via bactieria/pests **[1]**; Poor standards of food hygiene in people handling food can transmit food poisioning to themselves and to others **[1]**; Dust and soil can carry food poisioning onto or into food **[1]**

Buying and Storing Food
1. **a)** Water boils at **100 °C**. **[1]**
   b) The temperature of a freezer should be below **−18 °C**. **[1]**
   c) Bacteria multiply rapidly between **5 °C** and **63 °C**. **[1]**
   d) When heating food, the internal temperature should reach **75 °C**. **[1]**
2. **a)** False **[1]**
   b) False **[1]**
   c) True **[1]**
3. **Any four from**: Ensure the food is cooled before refrigerating or freezing it as the microorganisms will continue to multiply. If the food is hot, then it heats the fridge or freezer and therefore puts all the food within the danger zone **[1]**; Ensure the

food is kept in the refrigerator (between 0 and 5 °C) so that it is out of the 'danger zone' where microorganisms are most active **[1]**; Cool food quickly as this limits microorganism growth time, ensuring less time in the 'danger zone' **[1]**; Do not leave food out for long periods of time as this encourages the growth of microorganisms **[1]**; Keep food out of the 'danger zone' of 5–63 °C because this is where the microorganisms are most active – out of this range the microorganisms remain inactivated **[1]**; Use the food within a few days of cooking/opening because exposure to air and temperatures once open encourages food to deteriorate and microorganisms to multiply **[1]**; Transfer food to containers from tins and cans because open cans encourage microorganism activity and can lead to poisoning from the can oxidising **[1]**.

Preparing and Cooking Food
1. **a)** 5 °C **[1]**
   b) 75 °C **[1]**
   c) 5 °C **[1]**; 63 °C **[1]**
2. Reheating cooked foods: 75 °C (**accept 72 °C–75 °C**) **[1]**
   Storing chilled foods: blast chilling 4 °C is within range (**accept** 0–4 °C **or** 'below 5 °C') **[1]**
3. A food or temperature probe **[1]** (**'thermometer' will not be accepted**)
4. Sterilise/clean with antibacterial wipe (**clean not acceptable unless qualified, 'disinfect' will not be accepted**)/make sure sterile before and after use **[1]**; Check reading before start (ideally room temp), do not touch container/baking tin with probe, then place into centre of food/core temperature **[1]**; Keep in place for two minutes **[1]**; Check temperature is 75 °C or over **[1]**; Remove from food/take the reading when it is stable/after two minutes **[1]**; If not at correct temperature continue cooking food and re-apply later **[1]**.

Food Choices
1. **c)** **[1]**
2. an epi-pen **[1]**
3. Iron **[1]**; vitamin D **[1]**; Vitamin B12 **[1]**
4. Vegan, **any three from**: Peas (any type) **[1]**; Beans (any type) **[1]**; Lentils (any type) **[1]**; Nuts (any type) **[1]**; Seeds (any type) **[1]**; Soya products **[1]**.
   Lacto: Milk **[1]**; Cheese **[1]**; + any vegan food **[1]**.
   Ovo-lacto: **any three from**: Eggs **[1]**; Milk **[1]**; Cheese **[1]**; Quorn **[1]** + any vegan food **[1]**

British and International Cuisines
1. **b)** **[1]**
2. **b)** **[1]**
3. Own colour **[1]**; Own texture **[1]**; Own flavour **[1]**; Ingredients from the region that produces the cheese **[1]**
4. Protein: Any choice of meat (or vegetarian equivalent) **[1]**
   Starchy food: Roast or creamed potatoes **[1]**; Yorkshire pudding **[1]**

Vegetables, **any three from**: Carrots **[1]**; Broccoli **[1]**; Peas **[1]**; Any other suitable vegetable **[1]**
Sauce: Gravy **[1]**
Accompaniment, **any one of**: Mint (with lamb) **[1]**; Apple (with pork) **[1]**; Horseradish (with beef) **[1]**

Sensory Evaluation
1. **b)** **[1]**
2. **c)** **[1]**
3. **a)** Star profile/Radar diagram **[1]**
   b) Pizza base texture **[1]**
   c) Taste of cheese **[1]**; Golden brown cheese **[1]**
   d) Use stronger cheese **[1]**; to improve the taste of the pizza **[1]**
   More cheese when cooked **[1]**; would improve both colour and flavour **[1]**
   Cook pizza for longer **[1]**; until the cheese browns in the oven **[1]**.

Food Labelling
1. **c)** **[1]**
2. **b)** **[1]**
3. **a)** **Any six from**: Traffic light labelling educates consumers so that they can make informed choices **[1]**; Traffic light labelling identifies the nutritional content levels of the food **[1]**; Traffic light labelling is easy to read/interpret **[1]**; Colours may be explained: Red = poor choice for healthy eating **[1]**; Amber = caution in quantities eaten **[1]**; Green = free to eat in high quantities **[1]**; Should aim for more green, less red, and moderate amounts of amber foods **[1]**.
   b) **Any three from**: Energy value in kilojoules **[1]**; Energy value in kilocalories **[1]**; Amount in grams of fat **[1]**; Amount in grams of saturates **[1]**; Amount in grams of sugars **[1]**; Amount in grams of protein **[1]**; Amount in grams of salt **[1]**.
   c) **Any three from**: When the supplier makes a nutrition claim for the food product **[1]**; When the supplier makes a health claim for the food product **[1]**; If vitamins are added to the food product **[1]**; If minerals are added to the food product **[1]**

Factors Affecting Food Choice
1. **c)** **[1]**
2. **a)** The Eatwell Guide **[1]**
   b) What to cook, **any two from**: What do people in the family like? **[1]**; What ingredients are needed? **[1]**; How many people are you catering for? **[1]**
   How to cook it, **any two from**: Equipment available **[1]**; Skill needed **[1]**; Recipe **[1]**
   Time, **any two from**: When the food has to be served **[1]**; Who is eating and when **[1]**; How long you have to cook the food **[1]**

Page 95 Quick Test
1. Food miles are the distance that food travels from its origin to your plate

2. Recycling can use more energy to produce new packaging and some solutions used to clean recycled products are polluting

3. **Any two from**: Restrict catch sizes; Have minimum sizes of fish allowed to be kept for sale; Widen the selection of fish being eaten to more species; Put caught young fish back in the sea so that they can go on to breed and reproduce.

Page 97 Quick Test

1. So that if a risk is identified, the source can be quickly isolated and consumers can be protected

2. Animals have access to the outdoors as they would in nature; They can behave and feed naturally; The resulting product is better quality as the animal has not suffered stress

3. Hydroponic farming requires large areas of glasshouses, with controlled systems where the products grow in nutrient-rich liquids, without soil

Page 99 Quick Test

1. The Red Tractor logo flag shows the country of origin of the food product

2. Fairtrade products give a realistic income to the farmer; Fair prices for the goods; Local investment; Better working conditions

3. To make way for growing crops and animal grazing

4. Methane and CO_2

Page 101 Quick Test

1. Bran

2. Milk that has been processed to distribute the fat content throughout

3. **Any two from**: Single; Whipping; Double; Extra thick; Crème fraiche; Soured; Clotted

4. Durum wheat flour

Page 103 Quick Test

1. Vitamins A, C, E, and K

2. Six months

3. Vacuum has oxygen removed and MAP has oxygen removed and replaced with another gas

Pages 104-107 Review Questions

Food Choices

1. d) [1]
2. Anaphylaxis (or Anaphylactic shock) [1]
3. Lactose is a sugar [1]
4. a) People over 40 [1]; People who are overweight [1]
   b) Healthy diet [1]; Increased physical activity [1]
5. The sufferer would experience severe stomach pain [1]; The sufferer would experience anaemia [1]; The sufferer would eventually experience malnutrition [1]
6. Blood sugar level [1]

British and International Cuisines

1. b) [1]
2. Multicultural: due to the influence of people originally from other countries (or previous generations of their families) now living in the UK [1]
   Foods available, **any one of**: Specialist food shops that cater to particular cuisines or ethnic groups increase the diversity of food available [1]; International foods are now

available to buy online and in supermarkets [1]; Recipe books that expain how to make food from all over the world are readily available and relatively cheap to buy from bookshops of online [1]; Free recipes from all over the world are available online [1]. Lifestyle, **any one of**: People now take routinely take holidays to other countries, where they are exposed to different cuisines – they may take the new ingredients and recipes home with them [1]; We have a diverse range of takeaways/restaurants that sell the cuisines of other countries [1].

Sensory Evaluation

1. a) [1]
2. **Any three from**: Sweetness [1]; Flavour [1]; Colour [1]; Texture [1]; Appearance [1].
3. a) Ranking test [1]
   b) Triangle test [1]
4. So that the best aspects of the product can be retained [1]; So that the worst aspects of the product can be changed [1].

Food Labelling

1. a) 45 g [1]
   b) 1570 kJ/375 kcal [1]
   c) Handful/portion [1]; of fruit [1]
   d) Fibre: 3.7 g [1]; Saturates: 0.1 g [1]
2. **Any four from**: Allergen information [1]; Date mark [1]; Cooking instructions [1]; Storage instructions [1]; Batch mark [1].
3. b) [1]; d) [1]

Factors Affecting Food Choice

1. What the food looks like [1]; What the food smells like [1]; What the food tastes like [1]; What the food feels like in the mouth [1]
2. a) [1]
3. Buying foods in season: In-season food has more flavour [1]; In-season food has more nutritional value [1]
   Choosing where to shop, **any two from**: Type of shops available [1]; Income available [1]; Access – transport [1]
   Cost of food, **any two from**: What income is available [1]; Quality v cost [1]; Cost of seasonal food [1]; Taking advantage of special offers [1]
   Availability of different foods to buy, **any two from**: Multi-cultural influences [1]; Organic food choices [1]; Gluten-free options [1]
4. b) [1]

Pages 108-111 Practice Questions

Food and the Environment

1. a) carbon dioxide [1]
   b) your table [1]
   c) more [1]
   d) front [1]
2. Packaging that rots naturally [1]; Packaging that rots quickly [1].
3. The body gets delivery of the correct nutrients for that time of year [1]; also trace elements and minerals for the correct time of year [1].
4. You only pick what you need for that meal so there is less waste [1]; You have a variety of seasonal foods at the right time of year for your body's needs [1]; It is cost-efficient [1].

5. The peak harvest time for crops in the UK/the time of year when there is most demand for a particular crop [1]

Food Provenance and Production Methods

1. c) [1]
2. Less methane being produced by animals [1]; More land available to grow high-protein crops, e.g. soya [1].
3. low cost [1]
4. **Any three from**: Grown naturally [1]; No chemical or synthetic treatments [1]; No antibiotics [1]; Natural manure and compost only as fertilisers [1]; Must be GM-free [1].

Sustainability of Food

1. Growing different crops [1]; in a field each year to reduce soil erosion [1]; and improve general health of crops [1].
2. c) [1]
3. Rainforest [1]; has been cut down [1]; to grow and harvest palm trees on the land [1]. The trees in the rainforest remove CO_2 from the atmosphere [1]. Also, the rainforest is the habitat of many animals now facing extinction [1].
4. **Any four from**: Farmers in developing countries get realistic incomes [1]; Investment in local community [1]; Better working conditions [1]; Fair price for goods [1]; Sustainable production [1].

Food Production

1. Fresh milk has cream on top [1]; Homogenised has the cream dispersed throughout the milk [1].
2. Ultra Heat Treated
3. condensed milk
4. The changing or converting of raw food materials into, e.g. food that can be eaten immediately or be processed into other food products (wheat into flour) [1].
5. The changing or conversion of primary processed foods into other food products, which may involve combining one or more food ingredients, e.g. flour into pasta [1].

Food Processing

1. **Any three from**: Canning [1]; Irradiation [1]; Pasteurisation [1]; Sterilisation [1]; Ultra heat treatment [1].
2. a) Foods are irradiated to delay ripening.[1]
   b) Any two fruits or vegetables named that are imported from other countries, e.g. strawberries, bananas [2].
3. Accelerated Freeze Drying [1]
4. Generally there is no change to the food or nutrient content [1].
5. a) Sunlight drying, which allows moisture to evaporate from the food in the sunshine [1].
   b) **Any one of**: Fish, e.g. cod [1]; Meat, e.g. beef jerky [1]; Fruit, e.g. grapes– raisins [1].
   c) **Any three from**: Can affect the natural colour [1]; May develop a wrinkly surface [1]; Texture change [1]; Loss of vitamin C and B6, as well as others [1].
6. a) Modified Atmospheric Packaging [1]
   b) Sealed packages have oxygen removed and the gas content inside changed [1].

c) **Any two from**: Ready-prepared salads [1]; Vegetables [1]; Fruits [1].
d) Longer shelf life [1]; Fresh appearance [1]; Fresh taste [1].

Food and the Environment

1. a) Some of the chemicals used to clean during **recycling** [1] can cause **pollution** [1].
   b) Recycling may use more **energy** [1] than making packaging from new resources.
   c) **Food waste** [1] can be reduced by wise shopping and planning ahead.
   d) Food in landfill sites gives off methane gas as it **decomposes** [1].
   e) Recycling is a **green** [1] way of using collection facilities and bottle banks.
2. Restriction on the sizes of catches allowed [1]; Ensuring that fish caught must be returned if they don't meet a minimum size requirement [1]; Encouraging the public to eat a wider variety of fish [1].
3. A site where rubbish is dumped and buried [1].

Food Provenance and Production Methods

1. a) **Intensive farming** [1] produces large-scale low-cost food in sealed buildings, whereas **barn-reared** [1] animals have access to natural light and some environmental enrichments.
   b) **Organic** [1] grown foods are farmed naturally without any chemical or synthetic treatments and are **GM-free** [1]. This is a **lifestyle** [1] choice. **Free-range** [1] farming allows animals to go outdoors for part of their lives.
2. The ability to track any food [1;] animal feed [1]; food-producing animal [1]; or substance [1]; that will be used for consumption through all stages of production, processing and distribution [1].
3. No, it is very expensive [1]
4. Organic/free range [1]
5. They strip the seabed bare of every living creature for miles [1].

Sustainability of Food

1. a) c) e) f) [4]
2. True [1]
3. Large areas of forest being cut down in order to graze animals or grow crops [1].
4. **Any one of**: CO_2 builds up, which contributes to global warming [1]; Animals lose their natural habitat, putting them in danger [1]; Deforested land suffers from soil erosion and becomes infertile [1].

Food Production

1. Secondary [1]
2. Canned milk [1]; Dried mik [1].
3. Bacteria culture is added to milk. [1]
4. **Any three from**: Single [1]; Double cream [1]; Extra thick [1]; Soured [1]; Clotted [1]; Crème fraiche [1].
5. a) **Any one of**: Pouring over desserts [1]; Sauces [1].
   b) **Any one of**: Serving with ice cream [1]; Scones (clotted cream tea) [1].
   c) In place of cream in recipes (as it has a lower fat content) [1].

6. During processing the protein content of the milk coagulates and shrinks [1].

Food Processing

1. Vitamin B6 [1]; Vitamin C [1]
2. b) d) [2]
3. Usually no change to food or nutrient content [1].
4. a) **Any three from**: Vinegar [1]; Sugar [1]; Salt-brine solution [1]; Smoking [1]; Alcohol [1]; Vacuum packing [1]; MAP [1].
   b) **Any three from**: Vinegar – in pickling [1]; Sugar – in jam-making [1]; Salt-brine solution – for fish and meat [1]; Smoking – food cooked by heat from wood fires [1]; Alcohol – fruit stored in brandy [1]; Vacuum packing – oxygen is removed from the sealed package [1]; MAP – sealed package has gas content inside changed [1].
   c) All preserving methods are used to prolong shelf life [1].
5. Pickled in vinegar to make, for example, chutney/pickled onions/tomato sauce [1]; Both could also be chopped and frozen [1].
6. **Any two from**: Smoking [1]; Vacuum packaging [1]; Salt [1].

1. D [1]
2. Reference intake [1]
3. **Any three from**: Used for energy [1]; Helps insulate the body/keep warm [1]; Provides the body with fat-soluble vitamins A, D, E and K [1]; A component of hormones [1].
4. **Any three from**: a coarse texture results [1]; Over-rise takes place [1]; Collapse/sink in product's structure [1]; Unpleasant taste [1].
5. a) Heat is transferred from the hob through the saucepan by conduction [1].
   b) Heat is transferred through the milk by convection [1].
   c) Stirring (agitating) the milk distributes the heat throughout the milk [1]; preventing lumps [1].
6. **Any two from**: Babies/children [1] for growth [1]; Adolescents [1] for growth spurts [1]; Pregnant women [1] for the growing baby [1]; Nursing mothers [1] for lactation (milk production) [1].
7. Durum wheat; Stronger [1] **[both answers needed for 1 mark]**
8. B [1]
9. **Any three from**: Condensed milk [1] in tins [1]; Evaporated milk [1] in tins [1]; Dried milk [1] in card containers [1]; UHT milk [1] in card cartons [1].
10. a) Coagulation is the denaturation of protein [1]; It can be caused by heat during cooking [1]; It makes a quiche filling set [1].
    b) Any **one** example of cooked egg: Cakes [1]; Puddings [1]; Omelette [1].
11. a) Wheat flour thickens [1]; gelatinises [1]; during the cooking [1]; of the milk-based sauce.
    b) Rapeseed oil makes the potato starch [1]; turn golden brown [1]; during high temperature roasting [1].

c) Egg wash makes the pastry golden and shiny [1]; as the proteins set (coagulate) [1]; during baking [1].
d) Custard powder is cornflour, vanilla and yellow colour – so it thickens [1]; flavours [1]; and colours [1]; the milk.
12. They may be tanks on land or nets floating in seawater or fresh water [1]; Fish stocks are controlled, but still within a semi-natural environment [1].
13. a) 1 and 3 must be ticked. [1]
    b) Charred/blackened/burnt buns would indicate a colour change [1]; Charred/blackened/burnt buns would give a bitter taste [1]; Overbaked buns would be hard and unpleasant to eat [1].
14. Pot of yoghurt [1]
15. D [1]
16. a) Shortening [1]
    b) Emulsification [1]
    c) Gloss/flavour [1]
    d) Enriching [1]
    e) Aerating [1]
17. C [1]
18. **Any five from**: Large farms [1]; Large numbers of animals in crowded buildings [1]; Massive buildings [1]; No access to natural resources for the animals, e.g. daylight [1]; Very few farm workers needed to run computerised feeding systems [1]; Antibiotics used on the animals [1]; Growth enhancers used on the animals [1].
19. Conditions required are: Food – from sugar or flour [1]; Time – to allow doubling of dough size [1]; Moisture – the liquid in the recipe [1]; Warmth – 20–25 °C [1].
20. B [1]; C [1]; D [1]
21. a) 100 g [1]
    b) Green [1]
22. B [1]
23. Coronary Heart Disease [1]; Diabetes [1]
24. B [1]
25. In the following order: False [1]; True [1]; False [1]
26. Gluten [1]
27. **Any three from**: Warm temperature/37 °C [1]; Moisture [1]; Food [1]; Time [1]; Neutral pH [1]; May need oxygen [1].
28. A [1]; B [1]
29. B [1]
30. a) Remove the head [1]
    b) Fileting/Cook's knife [1]; Blue board [1]
    c) **Any three from**: Eyes are bright and not dull [1]; Scales are in place [1]; Gills are bright red [1]; Slightly salty smell of the sea [1]; Thin layer of slime [1]; Flesh is firm [1].
31. 1 Cornish pasty [1]; 2 Double Gloucester cheese [1]; 3 Eccles cakes [1]
32. **Any two from**: To make food passing through digestive system soft and bulky [1]; Helps prevent constipation [1]; Helps prevent diverticular disease and some forms of cancer [1].
33. C [1]
34. **Any one of**: Bread/or named bread [1]; Yoghurt/or named yoghurt [1].
35. **Any three from**: Making jams/jellies/crystallising [1]; Pickling/chutneys [1]; Bottling [1]; Drying [1]; Freezing [1]; Salting [1]; Smoking [1].

36. Vegetarians eat fish: False **[1]**; Hindus eat pork: False **[1]**; Jewish people don't eat shellfish: True **[1]**

37. cholesterol **[1]**

38. Sugar **[1]**

39. B **[1]**

40. a) **Any four from**: Breast **[1]**; Leg **[1]**; Wing **[1]**; Drumstick **[1]**; Thigh **[1]**
    b) Red **[1]**
    c) Boning knife/Cook's knife **[1]**
    d) Remove legs **[1]**; Separate the drumstick and the thigh **[1]**; Remove wings **[1]**; Find the wishbone at the front of the bird, release the wishbone **[1]**; cut through the knuckle at the base, then carefully remove the breasts from the carcass **[1]**.

41. B **[1]**

42. **Any two from**: Follow the Eatwell Guide **[1]**; Eat smaller portions **[1]**; Drink plenty of water **[1]**; Make sure that your diet contains plenty of calcium-rich foods to help prevent osteoporosis **[1]**.

43. True statements: Vegetarians may lack iron in their diet **[1]**; Lacto vegetarians eat honey **[1]**.

44. **Any two from**: Pigs – pork **[2]**; Cattle – beef **[2]**; Sheep – lamb **[2]**; Deer – vension **[2]**.

45. A **[1]**

46. C **[1]**

47. Food miles are the distance food travels from where it is produced to where it is eaten (also known as 'field-to-fork') **[1]**.

48. **Any three from**: To avoid cross contamination **[1]**; To keep raw foods and cooked foods apart **[1]**; To avoid bacteria transfer **[1]**; Reference to named bacteria **[1 mark per bacteria, up to three bacteria]**; Examples – chopping boards/knives/storage containers can be colour coded **[1]**; Colours used to identify equipment – Raw meat – red/Salad and fruit – green/Fish – blue/ vegetables – brown/Cooked meat – yellow/ Bakery and dairy – white **[1]**.

49. Global warming is caused by CO_2 emissions trapping the Sun's energy **[1]**.

50. **Any three from**: Table sugar (sucrose) **[1]**; Honey **[1]**; Jam **[1]**; Fruit juice **[1]**; Sweets **[1]**; Chocolate **[1]**; Cakes **[1]**; Biscuits **[1]**; Sugary drinks **[1]**.

51. High biological value (HBV) protein, **any three from**: Meat **[1]**; Fish **[1]**; Eggs **[1]**; Cheese **[1]**; Milk **[1]**; Soya **[1]**. Low biological value (LBV) protein, **any three from**: Nuts **[1]**; Seeds **[1]**; Lentils **[1]**; Pulse vegetables **[1]**; Bread **[1]**.

52. **Any eight from**: Identification of chicken as a high-risk ingredient **[1]**; Identification of eggs as a high-risk ingredient **[1]**; Identification of cream as a high-risk ingredient **[1]**; Identification of rice as a high-risk ingredient **[1]**; Cross contamination should be avoided **[1]**; Correct storage temperatures should be observed **[1]**; Food hygiene and cleaning procedures must be applied to hands, hair, etc. and equipment **[1]**; Correct reheating temperatures must be observed **[1]**; Correct cooking temperatures must be achieved **[1]**; Storage after cooking should meet the correct standards **[1]**.

53. Diabetes **[1]**; sugar **[1]**; insulin **[1]**

54. Enriched bread dough: the addition of sugar, butter and maybe egg **[1]**. Strong elastic dough: stretchy dough that does not break and shrinks back when pulled out of shape **[1]**. Rolling boil: Water that has been heated to the point where it has large bubbles that break on the surface **[1]**. A dough: A mixture that contains flour, and possibly other ingredients, held together by a liquid **[1]**.

55. a) The Red Tractor logo (controlled and monitored by Assured Food Standards (AFS)) **[1]**
    b) On food labels or on food packaging **[1]**
    c) **Any eight from**: The Red Tractor logo tells us that the food has been produced **[1]**, processed **[1]** and packed **[1]** to Red Tractor standards; The flag on the logo shows the country of origin **[1]**; The logo assures high standards of food hygiene and safety **[1]**; The logo assures high standards of equipment used in production **[1]**; The logo assures high standards of animal health and welfare **[1]**; The logo assures high standards in environmental issues and the responsible use of pesticides **[1]**; Any product with this logo can be traced from farm to fork **[1]**.

56. **Any two from**: In hot weather to replace water lost by excess sweating **[1]**; After physical exertion to replace water lost due to excess sweating **[1]**; Lactating mothers need an increased supply of water for the production of milk **[1]**; People suffering from sickness and diarrhoea need to increase their intake of water due to water loss through sweating/vomiting/excess excretion **[1]**.

57. C **[1]**

58. **Any two from**: Maintenance of normal iron metabolism **[1]**; Maintenance of normal vision **[1]**; Maintenance of skin and the mucus membranes **[1]**.

59. Moulds **[1]**; Yeast **[1]**; Bacteria **[1]**

60. **Any one of**: Use wholemeal or granary flour **[1]**; Use seeds as a topping **[1]**.

61. **Any six from**: Check the ingredients are fresh/within their sell-by date **[1]**; Do not use the same equipment to cut up the raw chicken and the vegetables **[1]**; Wash your hands before and after handling the raw chicken **[1]**; Use colour-coded equipment to avoid cross contamination **[1]**; Use the correct storage temperatures **[1]**; Ensure that the chicken is refrigerated until just before use **[1]**; Wash the chopping boards before use **[1]**; Follow food hygiene and cleaning procedures **[1]**; Use the correct reheating temperatures **[1]**; Use the correct cooking temperatures **[1]**; Make sure that the core temperature of the chicken reaches 70–72 °C (piping hot) **[1]**; Check that the chicken is cooked, e.g. cut it to check that the meat is not pink inside **[1]**; Cook the chicken prior to adding the vegetables **[1]**; Follow the correct storage procedures after cooking (cool quickly, cover properly, store in a fridge) **[1]**.

62. Roast chicken **[1]**

63. Calcium is added to white bread by law (fortified) **[1]**

64. Method: Roller drying **[1]** Details, spray drying: This method is used for some foods that are damaged by dry heats **[1]**. Details: Food is frozen **[1]** and the temperature is then increased to make the ice vaporise **[1]**.

65. Carbon dioxide **[1]**

66. **1–2 marks from**: Refrigerator temperature of 10 °C is too high/incorrect **[1]**; This temperature is in the danger zone/5–63 °C **[1]** **[extended answer]**. **Any further 4 marks from**: Effects on food stored at 10 °C: Perishable food will have its shelf life reduced/food will not last as long **[1]**; Bacteria will grow/breed **[1]**; Chilled food will not be safe to eat **[1]**; Danger of food poisoning **[1]**; Health concerns if food is eaten **[1]**; Quality of food will suffer, e.g. texture **[1]**; Correct temperatures should be: 1–60 °C for perishables **[1]**; Correct temperature range for high-risk perishables is 1–4 °C **[1]**.

67. Antioxidants protect us from pollutants in the environment **[1]**.

68. Eggs – in the door **[1]**; Mango – in the salad drawer **[1]**; Raw fish – on the bottom shelf or below the chocolate éclairs **[1]** **[the raw fish must be below the chocolate éclairs, fish in the salad drawer will not be accepted]**; Chocolate eclairs – on the first shelf or second shelf **[1]** **[must be above the raw fish]**.

69. **Any one of**: Sodium regulates the amount of water in the body **[1]**; Sodium assists the body in the use of energy **[1]**; Sodium helps to control muscles and nerves **[1]**.

70. Vitamin D **[1]**

71. C **[1]**

72. **Any six from**: Water in food/stored here remains frozen/solid **[1]**; Fast freezing produces small ice crystals so reduces damage to food structure **[1]**; Sensory properties, e.g. taste, flavour, colour, shape are maintained for most foods (not some fruit/veg) **[1]**; Nutritional value is maintained **[1]**; Food does not decay while correct temperature and storage time are maintained **[1]**; Shelf life of food is extended **[1]**; Bacteria stop growing – become dormant – water changes to ice crystals, making it unavailable for bacterial growth **[1]**; Bacteria not killed – will become active again on thawing **[1]**; Enzyme action is slowed down but not destroyed **[1]**.

73. Iron, main function in the body: Manufacture of haemoglobin in red blood cells **[1]**. Iron, main food sources: Red meat, especially liver **[1]**. Calcium, main function in the body: Strengthening of bones and teeth **[1]**. Calcium, main food sources: Milk and dairy products **[1]**.

Glossary

A

aeration processes in food preparation that incorporate air to give light, open textures or assist rising

allergy conditions, such as itchy skin or restricted breathing, caused by sensitivity to something in the environment or to a food

allergen a food that causes a bad chemical reaction

amino acids small molecules that form long chains in proteins

anaemia if there is a deficiency of iron in the blood, it is called anaemia; symptoms include tiredness

anaphylaxis an acute allergic reaction to a food, e.g. nuts, that can, in extreme cases, lead to death

antioxidants these help to protect cell membranes and to maintain healthy skin and eyes; vitamin E is an antioxidant

aroma a distinctive (usually pleasant) smell of a food

ascorbic acid another name for vitamin C

B

bacteria single-celled organisms that are able to reproduce rapidly (also called microorganisms); bacteria are sometimes useful and are used to help make food, e.g. cheese, and are sometimes harmful (pathogenic) and can cause food poisoning

barn-reared animals animals bred and housed in a large barn, rather than fully outdoors or in cages

Basal Metabolic Rate (BMR) the energy needed by the body to power your internal organs when completely at rest

batch the quantity of food products made in one operation, e.g. scones, bread rolls

batonnet ingredients cut into square-shaped lengths

best-before date a safety date found on foods and their packaging; best-before dates usually appear on foods that have a long shelf life such as canned, dried and frozen food products

binding various ingredients, such as water, egg, flour or breadcrumbs, are used to 'bind' other ingredients together

Body Mass Index (BMI) a measure that adults can use to see if they are a healthy weight for their height; the ideal healthy BMI is between 18.5 and 25

bridge hold a knife skill technique where the food being cut is secured with one hand held in the shape of a bridge, while the free hand slices beneath 'the bridge' with the knife

C

caramelisation change of colour and flavour when heating sugar or a sugar solution

carbon dioxide gas (CO_2) gas produced from yeast and chemical raising agents

carbon footprint the total amount of greenhouse gas emissions something causes – the 'something' can be an individual, a country, an event, a journey, etc.

cardiovascular anything that relates to the heart and blood vessels

chemical preservation any chemical that is added to a food product in order to extend its shelf life; chemical preservation usually inhibits the growth of bacteria

chemical raising agents food-safe chemicals that produce carbon dioxide gas, e.g. bicarbonate of soda, baking powder

chiffonade the name given to the technique of cutting leafy green vegetables into fine shreds

chilled food food that must be stored at a temperature of below 5°C

cholesterol a type of fat found in the blood, which is made in the liver and is also obtained from the food you eat; there is 'good' cholesterol (called HDL, high-density lipoprotein) that helps to break down fat and 'bad' cholesterol (called LDL, low-density lipoprotein) that clogs up blood vessels

chopping to cut something up using a knife

choux a type of pastry used for eclairs, choux buns and gougères

claw grip a knife skill technique where the food being cut is secured with one hand held in the shape of a claw, while the free hand slices the food with the knife

climate change a change in global or regional temperatures caused by increased levels of carbon dioxide due to the intensive use of fossil fuels such as coal and oil

coagulate protein denaturation by heat

coating a covering of one type of food product by another, e.g. fish coated in breadcrumbs

cobalamin another name for vitamin B12

coeliac coeliac disease is caused by the body's immune system mistaking substances in gluten (a protein found in wheat, barley, rye, and also oats, which contain a similar substance to gluten), as a threat and attacking them; Coeliacs cannot absorb nutrients if they eat gluten – this causes severe pain and can lead to anaemia and malnutrition

collagen the name of the main structural protein in connective tissue

colostrum the first milk produced by a lactating mother; it is full of antibodies that are useful to the new-born baby

combining to mix ingredients together, either by hand or with a machine such as a hand mixer

composting to make a compost out of decayed plant matter that can be used to fertilise new plants

condensed a liquid such as milk that has been evaporated, has not been sterilised, and has had sugar added to it

conduction heat transfer from one molecule to another

connective tissue tissues made of fibres that form a framework in the body to bind or separate other tissues and organs

constipation when faeces become difficult to expel from the body because they are hard and small

consistency mouthfeel sensory term related to thickness and smoothness of a mixture

contamination when a food product has become impure via contact with a harmful material, e.g. bacteria

convection heat transfer by rising and falling currents in a liquid or air

cooking the application of heat to change food from raw to cooked

core temperature the temperature at the centre of cooked food; reheated food should reach a core temperature of 75°C

creaming beating together yellow fat and caster sugar to aerate it

crop rotation the practice of growing different crops every year in the same field – switching the type of crop grown helps to keep the soil fertile

cross-contamination transferring bacteria from one place to another

crustacean crabs, lobsters, prawns, crayfish, shrimp and squid are all crustaceans

cuisine a style or method of cooking characteristic of a particular country, region or establishment

D

danger zone 5°C–63°C; the temperature zone that food must be kept out of because it is the zone in which bacteria will multiply and contaminate foods

dehydration the process of the removal of water

denaturation changing protein function by heat (cooking), acids, or mechanical actions

dextrinisation browning due to dry heat, charring food

diabetes   a medical condition caused by the inability of the pancreas to produce any, or enough, insulin to control the amount of sugar in the blood

dicing   to cut food into cubes

dietary fibre   a type of carbohydrate; it cannot be digested to provide energy

disaccharides   complex sugars that are formed when two monosaccharides join together

diverticular disease   when pouches form in the intestines that then become infected with bacteria

drying   to remove most or all of the moisture from a food; usually done to extend the shelf life of the food

E

Eatwell Guide   a guide that shows the proportions in which different groups of foods are needed in order to have a well-balanced and healthy diet.

elastin   a protein found in connective tissue that functions together with collagen; elastin can recoil like a spring and is found in parts of the body that have elasticity, e.g. the skin

emulsification   mixing liquids that do not normally mix, e.g. oil and water

energy density   the amount of energy calories (cal) or kilojoules (kJ) a food contains per gram

enriched dough   basic bread dough with the addition of sugar, butter and sometimes egg

enrobed   to cover a food stuff in a coating, e.g. breadcrumbs around fish

enzymes   biological catalysts found in all cells, usually made from protein; enzymes break down plant and animal tissues causing fruit to ripen, meat to tenderise and enzymic browning (also known as oxidation) to speed up

enzymic browning   browning of food due to enzymes in cut fruits and vegetables

essential amino acids   the eight amino acids that need to be provided in the diet for good health

Estimated Average Requirements (EARs)   tables used by nutritionists that provide guidelines to the energy needs of individuals at various stages of life

ethical   decisions or actions taken on the basis of strongly held moral beliefs or intellectual principles

evaporated   when a liquid is heated so that water is removed

F

Fair trade   an organisation dedicated to securing fair working, social and environmental conditions, sale prices and trade deals for farmers and manufacturers in developing countries

fat-soluble vitamins   these are vitamins A, D, E and K and they are stored in the body for longer periods than water-soluble vitamins

fatty acids   fat is made up of fatty acids and glycerol

field to fork   the name given to the food chain from the start of agricultural production to final consumption

filleting   a thin-bladed, flexible and very sharp knife used to fillet fish

fish farms   places where fish are commercially farmed

flexible   something that moves and bends easily

foam   the incorporation of a gas or air in a liquid, e.g. egg white foam

foetus   a baby still in the womb

food miles   the distances food travels from its point of origin to your table

food poisoning   illness caused by harmful bacteria (pathogens) multiplying in or on food; symptoms of food poisoning include stomach pains, diarrhoea, vomiting, nausea (feeling sick), and fever

food waste   food that is wasted by not being eaten or by being discarded; food can be discarded for many reasons, e.g. during production due to contamination or in supermarkets due to the expiration of use-by dates

fortified   strengthening the nutritional content of a food by adding vitamins and minerals

free range farming   farming where the animals reared can roam freely outdoors for a portion of the day rather than being confined in a small space indoors 24 hours a day

frozen food   food that has been frozen as quickly as possible after it has been harvested or made and is kept frozen until it is required to be cooked for consumption

G

gelatine   a water-soluble protein that comes from collagen and is used in food preparation; it is colourless and tasteless

gelatinisation   chemical term for thickening by starch and starchy foods

gliadin   one of the two main components of the gluten fraction of the wheat seed (the other component is glutenin)

gluten   a protein found in wheat flour

glutenin   one of the two main components of the gluten fraction of the wheat seed (the other component is gliadin)

glycerol   part of a fat molecule

goujons   strips of chicken or fish that are deep-fried

Green House Gas (GHG)   any of a number of gases that absorb solar gases that contribute to the greenhouse effect, which contributes to global warming and climate change; examples include carbon dioxide, fluorocarbons, and methane

Genetically Modified (GM)   foods that have had changes introduced into their DNA using genetic engineering; GM foods can be manipulated to emphasise certain traits or to extinguish negative characteristics

H

haemoglobin   a red protein (it gives blood its colour) that transports oxygen around the body via red blood cells

halal   meat that can be eaten by Muslims because it has been killed in accordance with Islamic law

hatcheries   buildings/pools where fish or poultry eggs are artificially controlled and then hatched

heavy dropping consistency   when a mixture drops heavily from a spoon, rather than flowing from the spoon in a liquid or sticking to it in a lump

High Biological Value (HBV)   food sources, such as animal proteins, that contain all the essential amino acids required by the body

high-risk foods   foods that have a high risk of harbouring harmful bacteria, e.g. raw meat, raw seafood, cooked rice

homogenised   milk that is forced through tiny holes in a machine; this breaks up the fat and disperses it, and it doesn't reform as a layer

hydrogenation   making solid fat from a liquid oil

hydroponics   the name given to the process of growing plants without soil but in sand, gravel or liquid with nutrients added to stimulate growth

hygiene   the practice of preventing contamination by following strict cleanliness procedures

I

ingredients   single food items within a recipe

intensive farming   commercial agriculture on an industrial scale designed to maximise yields and profit; this type of farming relies on heavy use of pesticides and chemical fertilisers

iron deficiency anaemia   a common condition in teenage girls who have started menstruating (the blood loss incurred by menstruation causes iron deficiency)

J

jardinière   ingredients cut into baton-shaped lengths

julienne   ingredients cut into matchstick-sized strips

K

kosher   food that conforms to Jewish dietary law

kwashiorkor   a severe form of protein malnutrition

L

lactating   the production of milk by women from the mammary glands to feed new-born babies (all mammals lactate)

lacto vegetarian   a person who eats only dairy foods, no eggs, meat or fish

lactose intolerance   a condition where someone is intolerant of lactose, which is a natural sugar found in milk; lactose intolerance causes stomach upset

landfill   waste disposed of via landfill is dumped into a massive hole in the ground and eventually buried

Low Biological Value (LBV)   protein from plant sources is of Low Biological Value (LBV) and lacks some essential amino acids; the exception is soy, which is a plant protein of HBV

M

macédoine   ingredients cut into medium-sized dice

Maillard reaction   the browning of meat, caused by a reaction with natural sugars and proteins which produces a dark colour; also known as non-enzymic browning

malnutrition   a significant lack of proper nutrition

manufacturer   a company producing large quantities of food for sale to the public

marinade   a mixture created to tenderise and flavour meat, fish and alternatives

menstruation   the monthly process the human female body goes through to discharge the lining of the uterus; takes place from puberty to menopause

microorganisms   another name for bacteria

microwave   using a type of radiation called microwaves (produced by a magnetron), which travel in straight lines and penetrate food to cook or reheat it

milling   the process whereby wheat is made into a flour; wheat grains are blended with other varieties and washed to remove grit and dirt then huge rotating rollers crush the grains at varying speeds – the crushed grain is sieved and this produces flour

mollusc   oysters, mussels, scallops, winkles and cockles are all molluscs

monosaccharides   the simplest form of carbohydrate structure

moulds   a type of fungus that can settle on food and grow into a visible plant; a sign of food that is not fresh

multicultural   foods from a variety of countries, regions and cultures

muscle fibre   cells that give structure to muscles; different structures of muscle fibre indicate different types of muscle

myoglobin   a protein found in most mammals that is oxygen-binding and protein-binding

N

non-enzymic browning   browning due to caramelisation or dextrinisation, not caused by enzymes

Non–Starch Polysaccharide (NSP)   the correct scientific name for fibre

nutrition labelling   labels that show which nutrients a food product contains

O

olfactory   relating to the sense of smell

organic   food that is grown or produced without using chemical fertilisers or pesticides

origin   where something first comes from

osteomalacia   the name of the medical condition where bones become soft due to a lack of calcium or vitamin D

osteoporosis   the name of a medical condition where bones become weak, brittle and break easily

ovo-lacto vegetarian   vegetarians who eat eggs and dairy products (but only cheese that has been made with vegetable rennet)

oxidation   changes in food such as colour change or flavour change due to exposure to oxygen

P

packaging   the items used to wrap around foodstuffs; designed to protect and preserve

palatability   to make food appealing through its appearance, colour, flavour, texture and smell

palate   a person's ability to be able to distinguish between, and appreciate, different flavours

paring   a small knife used to prepare fruit and vegetables

pasteurised   a process of part sterilisation conducted via heat treatment or irradiation; pasteurisation makes food safer to eat and lengthens its shelf life

pathogens   bacteria/microorganisms that cause disease

peak bone mass   when bones reach their maximum strength

pH level   the level of acidity or alkalinity of a food

photosynthesis   the process that plants use to convert energy from sunlight into chemical energy that the plant can use to feed itself

Physical Activity Level (PAL)   the energy needed by the body for movement of all types

physical raising methods   using air, steam or water vapour to lighten and raise mixtures

plasticity   how fat properties change over a range of temperatures

polysaccharides   polysaccharides are made up of many monosaccharide units joined together

precise   accurate

pre-packed   food that has been packed before the customer buys it

preservation   the process of altering a foodstuff so that it can be kept for as long as possible, e.g. canning, pickling

primary processing   the conversion of foods in their raw state into other food products by combining the raw ingredient with other foods

probiotic   probiotic cultures are carefully selected strains; there is good evidence that they help improve digestion, safeguard the immune system, and keep the body's intestinal flora in balance

proportions   the parts of a food stuff comparative to the whole amount

protein complementation   when the proteins of a LBV can be eaten together to provide all the essential amino acids, e.g. beans on toast

proving   the last stage of rising bread goes through prior to baking

puberty   the stage of life when adolescents mature and become capable of sexual reproduction

R

radiation   heat transfer method, a type of electromagnetic radiation that involves waves, e.g. grill

rating   assessing something in terms of quality, quantity or a combination of both

ratio   proportion of ingredients in a basic recipe

recycling   re-using/converting waste materials, including food

Red Tractor   a logo that tells the consumer that the food has been produced, processed and packed to Red Tractor standards; it assures standards of food hygiene and safety; equipment used in production; animal health and welfare; environmental issues and responsible use of pesticides

Reference Intake (RI)   the approximate amount of a nutrient provided by a portion of food

regional   belonging to one particular geographical region

reheating   to heat something again; reheated food should reach a core temperature of 75°C

rest   when a dough is left, usually in a fridge or similar cold location, for a period of time before it is baked

retailer   an individual or organisation that sells goods

riboflavin   another name for vitamin B2

rickets   a childhood disease caused by a lack of vitamin D; it causes softening of the bones, which results in bow legs

rolling boil   when something is boiling all over the surface (both middle and sides) with great energy

rubbing-in   incorporating fat into flour when making pastry

S

salting   the name given to the process of adding salt to a foodstuff to remove its moisture; food poisoning bacteria cannot survive without moisture

saturated fats   come mainly from animal food products, e.g. lard, and they are solid at room temperature; saturated fats raise your cholesterol level – too much bad cholesterol can lead to health problems

secondary processing   the processing of raw food ingredients into other food products

semi-skimmed   pasteurised milk that has had some of the cream removed

senses   perception of a food using hearing, sight, touch, smell and taste

sensory analysis   the application of sight, smell, taste, touch and hearing to evaluate foods

sensory characteristic   using the senses to evaluate food products, e.g. aroma, texture

serrated   a type of knife blade that is jagged; often used to carve joints of meat

shortcrust   a type of pastry used for pies and tarts

shortening   mouthfeel, such as short or crumbly, produced by short gluten strands

skimmed   pasteurised milk that has had all or most of the cream removed

smoking   the name given to the process of smoking a foodstuff to a temperature of 76 °C or above; this removes the moisture, extends shelf life, and imparts a distinctive flavour

spina bifida   a defect of the lower spine which can result in paralysis in the legs and feet and is sometimes accompanied by learning difficulties

starter culture   a mixture of flour and water that ferments with bacteria; a starter culture aids fermentation in bread making

sterilised   subjecting a food to a high temperature to kill any bacteria that may be present; typically heating to between 110 °C and 135 °C takes place for 15–20 minutes

sustainable food   food that will continue to be available for many years to come

T

taste   sensation of flavour in the mouth on coming into contact with food

temperature   a measure of intensity of heat or cold

texture   properties of a food sensed by mouthfeel

thiamin   another name for vitamin B1

thyroid   this is a gland at the front your neck, just below the Adam's apple; it produces two hormones which control the body's metabolic rate

traceability   knowing the history of a foodstuff from the place of its birth/planting all the way through its manufacturing journey to the consumer

trans fats   trans fats can be formed when oil goes through the process of hydrogenation to form a solid; this process occurs as the molecules flip and rotate

transportation   moving items or people from one place to another using various forms of transport, e.g. plane, lorry, boat

U

Ultra-Heat Treated (UHT)   the process of heating food above a temperature of 135 °C; this sterilises the food and lengthens its shelf life

unsaturated fats   come from plants, e.g. olive oil, and some animals; they are liquid at room temperature

use-by date   a safety date found on foods and their packaging – the food must be eaten by the date on the packaging; found on highly perishable packaged food, e.g. meat

V

vegan   a person who does not eat any animal or animal products

vegetarian   a person who chooses not to eat foods which involve killing animals or fish

viscosity   resistance to flow, thickness of a sauce or mixture

volume   the amount of liquid in a container, e.g. milk in a measuring jug

Y

yeast   a biological raising agent that ferments to produce carbon dioxide

Index

Aeration 47
Allergens 83
Allergies 77
Amino acids 20

Bacteria 58–59, 62–63
Baking powder 49
Basal Metabolic Rate (BMR) 31
Best-before dates 64
Bicarbonate of soda 49
Binding 13
Biological raising agents 49
Body Mass Index (BMI) 31
Bone health 30
Boning a chicken 11
Bowel cancer 30
Bridge hold 6
British cuisine 78
Buying food 64–65

Calcium 26
Caramelisation 45
Carbohydrates 22–23, 44–45
Carbon emissions 94, 98
Cheese production 60
Chemical raising agents 48–49
Chilled food storage 65
Cholesterol 21
Classifications of meat 10
Claw grip 6
Cleaning 67
Climate change 94, 98
Coating 13
Coeliac disease 77
Composting 95
Conduction 41
Constipation 23
Contamination 63
Convection 41
Cooking 40–41, 66
Cooking methods 41
Crop rotation 98
Cutting 6–7, 12

Dairy industry 60
Date marks 82–83
Defrosting 65
Dental health 30
Dextrinisation 45
Diabetes 30, 77
Dietary fibre 22–23
Disaccharides 22
Diverticular disease 23
Dough 14–15

Eatwell Guide 28
Emulsions 47
Energy 31
Enriched dough 14

Enzymes 42, 59
Enzymic browning 43, 59

Fairtrade 99
Farming 96–97
Fat 21, 46–47
Fat-soluble vitamins 21, 24
Filleting fish 9
Fish 8–9
Fish farming 97
Fluoride 26
Food labels 82–83
Food miles 94
Food poisoning 62–63, 67
Food preservation 102
Food waste 94–95
Food-borne disease 63
Free-range farming 97
Frozen food storage 65

Gelatinisation 44
Genetically Modified (GM) 96–97
Glazing 13
Gluten 14, 43

Healthy eating 84
Heart disease 30
Heat transfer 41
High-risk foods 66–67
Hydrogenated fat 21
Hydroponics 97
Hygiene 66

Intensive farming 96
Intolerances 77
Iodine 27
Iron 26
Iron-deficiency anaemia 29, 30

Japanese cuisine 79

Knives 7
Kwashiorkor 20

Laminated dough 14
Landfill 94–95
Lifestyle 85

Marinating 11
Meat 10–11
Meat industry 61
Meat structure 10
Micronutrients 26–27
Microorganisms 58–59, 60–61
Microwaves 40
Milk 100
Minerals 26–27
Modified starches 45
Monosaccharides 22
Moulds 59, 60

Non-starch polysaccharides (NSP) 23

Obesity 30
Oils 46–47
Organic foods 96–97
Osteoporosis 29, 30
Oxidation 43

Paired preference tests 81
Pasta 15
Pastry 14–15, 46
Physical Activity Level (PAL) 31, 84
Plasticity 46
Polysaccharides 22
Probiotic cultures 61
Protein 20, 42–43
Protein coagulation 42
Protein complementation 20
Protein denaturation 42
Puberty 29

Radiation 41
Raising agents 48–49
Ranking tests 81
Rating tests 80–81
Raw fish 9
Red Tractor 99
Reheating 67
Religion 76–77

Seasonality 85
Senses 80
Sensory analysis 80–81
Sensory evaluation 80–81
Sensory profiles 81
Shortening 46
Sodium 27
Spanish cuisine 79
Storing food 64–65
Sugar 22–23
Sugar substitutes 22
Sustainable food 95, 98–99

Temperature probes 67
Tenderising 11
Traceability 96
Trans fats 21
Triangle testing 81

Use-by dates 64

Vegetable cuts 6–7
Vegetarians 76
Vitamins 24–25

Water 27
Water-soluble vitamins 24–25
Wheat 101

Yeast 49, 58–59, 61
Yoghurt 61

Collins

AQA GCSE 9-1
Food Preparation and Nutrition

Workbook

Kath Callaghan, Fiona Balding,
Barbara Monks, Barbara Rathmill
and Suzanne Gray with Louise T. Davies

Contents

Food Preparation Skills

Knife Skills   148

Fish   149

Meat   150

Prepare, Combine and Shape   151

Dough   152

Food Nutrition and Health

Protein and Fat   153

Carbohydrate   154

Vitamins   155

Minerals and Water   156

Making Informed Choices   157

Diet, Nutrition and Health   158

Food Science

Cooking of Food, Heat Transfer and Selecting Appropriate Cooking Methods 159

Proteins and Enzymic Browning   160

Carbohydrates   161

Fats and Oils   162

Raising Agents   163

Food Safety

Microorganisms, Enzymes and Food Spoilage   164

Microorganisms in Food Production   165

Bacterial Contamination   166

Buying and Storing Food   167

Preparing and Cooking Food   168

Factors Affecting Food Choice

Food Choices   169

British and International Cuisines   170

Sensory Evaluation   171

Food Labelling   172

Factors Affecting Food Choice   173

Contents

Food Provenance

Food and the Environment **174**

Food Provenance and Production Methods **175**

Sustainability of Food **176**

Food Production **177**

Food Processing **178**

Practice Exam Paper 1 **179**

Practice Exam Paper 2 **196**

Answers **213**

1 Which type of knife is shown in the picture? Tick (✓) **one** answer.

a) Paring knife ☐

b) Cook's knife ☐

c) Filleting knife ☐

d) Palette knife ☐

[1]

2 When boning a chicken, what colour board should you use? Tick (✓) **one** answer.

a) Blue ☐

b) Red ☐

c) Green ☐

d) Brown ☐

[1]

3 A chef or professional cook has their own set of knives that are really the 'tools of the trade' and they learn to use them and look after them when training.

Name and describe the **two** main methods of cutting that are the basis of a cook's knife handling skills. [4]

	Knife Hold:
	Explanation:
	Knife Hold:
	Explanation:

4 Explain **four** safety rules to observe when working with knives in the kitchen.

[4]

Total Marks _____ / 10

Fish

1 Which of the following fish would you expect to find preserved in a can? Tick (✓) **one** answer.

   a) Cod ☐   **b)** Plaice ☐   **c)** Sardine ☐   **d)** Halibut ☐   [1]

2 Which of the following fish is a mollusc? Tick (✓) **one** answer.

   a) Oyster ☐   **b)** Crab ☐   **c)** Shrimp ☐   **d)** Lobster ☐   [1]

3 Which choice would NOT be suitable when enrobing a fish fillet before cooking?
Tick (✓) **one** answer.

   a) Breadcrumbs ☐   **b)** Polenta ☐

   c) Batter ☐   **d)** Chocolate ☐   [1]

4 In what **two** ways can fish be protected when cooked in hot fat?

   .... [2]

5 Complete the following statements using the words in the boxes below.

| collagen | 60 °C | coagulate | muscle | gelatine | connective tissue |

   Fish cooks quickly because the .... is short and the

   .... is thin.

   The connective tissue is made up of .... and will change into

   .... and .... at

   ....   [6]

6 The picture below shows a fish being filleted. Complete the table below.

	Classification of the fish shown: .... [1] **One** example of this type of fish: .... [1]	What is the chef showing? ....[2]

Total Marks .... / 15

Food Preparation Skills

Meat

1 Which of these nutrients is **not** found in meat? Tick (✓) **one** answer.

a) Fat ☐   b) Vitamin B6 ☐   c) Protein ☐   d) Vitamin C ☐   [1]

2 Which meat is classed as offal? Tick (✓) **one** answer.

a) Pork ☐   b) Poultry ☐   c) Liver ☐   d) Goose ☐   [1]

3 What is meat? Complete the following sentence using the words in the boxes below.

muscle	connective tissue	fibres

Meat is a _____ composed of cells consisting of

_____, held together by _____.   [3]

4 Name **one** part of an animal that is likely to be tougher to eat.

_____   [1]

5 Name **two** slow methods of cooking that are suitable for tough cuts of meat.

_____   [2]

6 Complete the following sentences about what happens when meat is cooked. Use the words in the boxes below.

Maillard	gelatine	coagulate	browning	sugars

The _____ of meat is caused by a reaction with natural _____ and

proteins to produce a dark colour. This occurrence is called the _____ reaction or

non-enzymic browning.

As the meat cooks the proteins _____ and produce a firm texture. Collagen is

broken down into _____.   [5]

Total Marks _____ / 13

Prepare, Combine and Shape

1 Which glaze would be most suitable for a batch of Chelsea buns? Tick (✓) **one** answer.

a) Egg wash ☐   b) Arrowroot ☐

c) Sugar and water ☐   d) Egg yolk ☐   [1]

2 What type of binding agent is generally used in a sausage mixture? Tick (✓) **one** answer.

a) Breadcrumbs ☐   b) Flour ☐

c) Egg ☐   d) Water ☐   [1]

3 In baking, what is the difference between stirring and whisking?

_____

_____ [2]

4 In baking, what is the advantage of using cutters and a rolling pin when making a batch of biscuits?

_____

_____ [2]

5 Explain which mixing and shaping skills can be used when making a decorated Victoria sandwich cake to give a quality finish.

_____

_____

_____

_____

_____

_____ [8]

6 How would you make sure that a batch of burgers are all exactly the same size and shape?

_____

_____ [2]

Total Marks _____ / 16

1 What is gluten? Tick (✓) **one** answer.

a) A leavening agent ☐   b) A sweetener ☐

c) A protein found in flour ☐   d) A muscle ☐   [1]

2 What is enriched dough? Tick (✓) **one** answer.

a) A dough that contains organic ingredients. ☐

b) A dough that has additional sugar and butter added. ☐

c) A dough that has extra ingredients added, such as herbs or cheese. ☐

d) A dough that contains expensive ingredients. ☐   [1]

3 What is the name of the method used to make shortcrust pastry?

.... [1]

4 What is the function of water in a shortcrust pastry?

.... [1]

5 Use the words in the boxes to complete the stages in the choux pastry method.

| cool | accurately | rolling boil | heavy dropping |

| fat | flour | water | paste |

Weigh all the ingredients ....

Place .... and .... in a pan and bring

to a ....

Add sieved .... immediately and mix well to form a

....

.... then add beaten eggs gradually to a

.... consistency. Cook in a hot oven.   [8]

6 What are the names of the **two** proteins found in strong plain flour used in making bread?

.... [2]

Total Marks .... / 14

1 Fill in the missing words.

The body needs protein for g.., maintenance and r... [2]

2 Give **two** different protein foods suitable for a lacto-vegetarian.

.... [2]

3 Give **three** examples of protein foods that have a High Biological Value (HBV).

.... [3]

4 Underline the food that is a good example of protein complementation.

Jam sandwich   **Lentil soup and bread**   **Tomato and basil salad**   **Sausage roll** [1]

5 What is the name of the protein deficiency disease? Tick (✓) **one** answer.

a) Beri-beri ☐   **b)** Scurvy ☐   **c)** Kwashiorkor ☐   **d)** Dermatitis ☐ [1]

6 Give **one** example of a fat in liquid form.

.... [1]

7 Which of the following are functions of fat in the diet? Underline **two** answers.

Provides concentrated energy   **Strengthens teeth**

Makes red blood cells   **Provides body with vitamins A, D, E and K**

Promotes growth [2]

8 Which **two** components is fat made up of?

.... [2]

9 What is hydrogenation?

....

.... [1]

Total Marks .... / 15

1 What are the **three** carbohydrate groups?

_____

_____ [3]

2 Give an example of a monosaccharide.

_____ [1]

3 Fill in the missing words.

Sugars are d _____ very quickly in the body, providing instant e _____ . [2]

4 Most people in the UK do not eat enough dietary fibre. Suggest a similar food that is higher in dietary fibre to replace each of those listed below.

a) White bread _____

b) Cornflakes _____

c) Mashed potato _____ [3]

5 Sugar, sweets and sugary drinks are associated with which type of decay in the body?

_____ [1]

6 What would be the results of not eating enough carbohydrate?

_____

_____ [2]

7 Fill in the missing words.

St_____ have to be digested into s_____ before

a_____ – this is s_____ e_____ release. [5]

8 What is the name of the common medical condition frequently caused by a lack of dietary fibre (NSP) in the diet?

_____ [1]

Total Marks _____ / 18

1 Name **two** dietary sources of vitamin A.

.... [2]

2 The B group vitamins and vitamin C (ascorbic acid) belong to which vitamin group?

.... [1]

3 What is rickets and what vitamin deficiency is it associated with?

....

.... [2]

4 A deficiency of vitamin B2 Riboflavin can cause what deficiency symptoms? <u>Underline</u> **one** answer.

Broken and split nails   **Sore throat**   **Vomiting**   **Skin cracking around the mouth**   [1]

5 One of the functions of vitamin C (ascorbic acid) is to act as an antioxidant. What do antioxidants do?

.... [1]

6 What is cholecalciferol?

....

.... [2]

7 Fill in the missing words.

When making a salad, to avoid the loss of vitamin C (a..

a..) through o.., prepare just before

s.. and avoid excess c.. .   [4]

8 Which vitamin deficiency can cause night blindness? Tick (✓) **one** answer.

a) Vitamin K   ☐   b) Vitamin A   ☐

c) Vitamin D   ☐   d) Vitamin B1   ☐   [1]

Total Marks .... / 14

1 Which vitamin helps the body to absorb calcium? Tick (✓) **one** answer.

a) Vitamin A ☐   b) Vitamin C ☐

c) Vitamin K ☐   d) Vitamin D ☐   [1]

2 Give **one** reason why calcium is needed in the body.

_____ [1]

3 Name a condition that an excess of sodium (salt) in the diet is linked to.

_____ [1]

4 What is salt mainly used for in food preparation?

_____ [1]

5 Fill in the missing words.

A lack of iron in the diet can cause iron-deficiency a_____ and symptoms include

t_____. [2]

6 What is the daily recommended intake of water?

_____ [1]

7 What is the function of fluoride in the diet?

_____

_____ [1]

8 Iodine supports the correct functioning of which gland in the body? Tick (✓) **one** answer.

a) Thyroid ☐   b) Pituitary ☐

c) Saliva ☐   d) Adrenal ☐   [1]

Total Marks _____ / 9

1 Name the **five** sections of the Eatwell Guide.

....

....

....   [5]

2 Fill in the missing words.

H............................ m............................ provides babies with all their nutritional

requirements, except for i............................. Babies are born with a supply of this stored in

their l............................   [3]

3 What is colostrum?

.... [1]

4 During pregnancy, why is a good supply of folate (folic acid) required?

....

.... [2]

5 Why are older people advised to eat lots of calcium-rich foods? Tick (✓) **one** answer.

a) To help prevent dementia ☐   b) To keep them hydrated ☐

c) To help strengthen bones ☐   d) To give them a better appetite ☐   [1]

6 Fill in the missing words.

The E............................ G............................ shows the proportions of food groups that

should be eaten daily for a w............................-b............................ diet.   [2]

7 What would you advise a pregnant woman who is suffering from constipation to do?

....

.... [2]

Total Marks .... / 16

1 Fill in the missing words.

C_____ H_____ D_____ is linked to diets high

in s_____ f_____, which make c_____ in

the blood. **[3]**

2 Tick (✓) **one** answer. Symptoms of iron deficiency anaemia include:

a) blurred vision and cataracts. ☐   b) sickness and diarrhoea. ☐

c) tiredness, weakness and lack of energy. ☐   d) a sore throat. ☐ **[1]**

3 Fruit and vegetables form part of a balanced diet. How many portions of fruit and vegetables are we advised to eat every day?

_____ **[1]**

4 List **three** different health problems that may be linked to a high consumption of fat in the diet.

_____

_____ **[3]**

5 Fill in the missing words.

In Type 2 diabetes, too little or no i_____ is produced, resulting in high levels of

s_____ in the b_____. **[3]**

6 What is osteoporosis? Tick (✓) **one** answer.

a) Stiff neck ☐   b) When bones become weak and break easily ☐

c) Weak connective tissues ☐   d) Persistent nose bleeds ☐ **[1]**

7 An individual's BMR depends on **three** things. What are these three things?

_____   _____   _____ **[3]**

Total Marks _____ / 15

1 What does heat transfer in liquids and in air cause? Tick (✓) **one** answer.

a) Halogen currents ☐

b) Conduction currents ☐

c) Convection currents ☐

d) Induction currents ☐ [1]

2 What is cooking?

....

....

.... [3]

3 Cooking methods can change the nutritional value of a food. Fill in the blanks to complete the chart.

Cooking Method	Change to Nutritional Content	Example
Dry method		Grilling sausages
Water-based (moist) method		

[3]

4 This question is about selecting appropriate cooking methods. Use the words in the boxes below to complete the sentences correctly.

| moist | hot | conserve | sensory | add | tough | quick |

Cooking methods can alter the .... properties of food by adding crispness or softening.

The right cooking methods can .... vitamins, or .... energy value.

When meat needs cooking, long, slow .... cooking is best for

.... cuts of meat.

If the meat is tender a .... and .... method of cooking such as grilling can be used. [7]

Proteins and Enzymic Browning

1 Tick (✓) **one** answer. The main function of eggs in a quiche tart is:

a) to aerate the filling. ☐

b) to add flavour. ☐

c) to set the filling. ☐

d) to reduce the number of calories. ☐ [1]

2 Food preparation before cooking, for example marinating, is a requirement of some recipes.

a) What is a marinade?

....

....

.... [3]

b) Explain why a marinade is used in cooking.

....

....

....

.... [3]

c) What might be the advantages of using a marinade for chicken kebabs?

....

....

....

.... [4]

d) Name **two** quick cooking methods that could be used to cook chicken kebabs.

.... [2]

3 Gluten is a protein found in wheat. Which statements about gluten are **true** and which are **false**? Circle one answer for each of the parts a)–d).

a) Gluten makes dough stretchy and elastic. **True/False** [1]

b) Gluten forms the structure of a baked loaf of bread. **True/False** [1]

c) Salt causes gluten to be weakened. **True/False** [1]

d) Gluten helps pasta hold its shape when cooked. **True/False** [1]

Total Marks .... / 17

1 Which one of the following is not a sauce making method? Tick (✓) **one** answer.

a) Creaming method ☐

b) All-in-one method ☐

c) Roux method ☐

d) Blending method ☐   [1]

2 Carbohydrates are useful functional ingredients. Which of the following is not a function of carbohydrate? Tick (✓) **one** answer.

a) Caramelisation ☐

b) Coagulation ☐

c) Dextrinisation ☐

d) Gelatinisation ☐   [1]

3 A cafe needs to make macaroni cheese for lunch service. The chef uses the ingredients listed below. Give **one** main function of each of these ingredients in the macaroni cheese.

semi-skimmed milk ....

wheat flour ....

margarine ....

seasoning ....

Red Leicester cheese ....

macaroni ....   [6]

4 What are the six stages of making macaroni cheese?

1 ....

2 ....

3 ....

4 ....

5 ....

6 ....   [6]

5 How does the macaroni cheese sauce become thick and smooth?

....

....

....

....   [5]

Total Marks _____ / 19

1 Tick (✓) **one** answer. The main function of fat in pastry is:

a) to combine other ingredients. ☐   b) to shorten the texture. ☐

c) to spread easily. ☐   d) to add bulk to the mix. ☐   [1]

2 This question is about understanding emulsified sauces.

> Basic recipe for Hollandaise sauce:
>
> 250 ml white wine vinegar
>
> 2 egg yolks
>
> 100 g melted butter
>
> Seasoning

a) Which **two** ingredients would not mix easily?

_____ [2]

b) What is the function of the seasoning? What might the seasoning be?

_____

_____ [2]

c) Explain the function of the egg yolk in the sauce.

_____

_____

_____ [3]

3 This question is about function of ingredients in pastry made with half fat to flour.

Choose the correct words from the boxes to complete the following sentences.

steam	dextrinisation	gluten	binds	shorten

a) Water _____ the ingredients together.   [1]

b) When pastry is cooked _____ of starch gives colour.   [1]

c) Fat is used to _____ pastry.   [1]

d) Water creates _____ to help pastry rise.   [1]

e) Rubbing in fat stops the development of long strands of _____ .   [1]

Total Marks _____ / 13

1 Why is water an effective raising agent? Tick (✓) **one** answer.

a) It turns to steam. ☐　　**b)** It does not add calories. ☐

c) It makes mixtures runny. ☐　　**d)** It makes mixtures moist. ☐　　[1]

2 Fill in the table by naming **two** chemical raising agents and giving an example of their use in food preparation.

Names of Chemical Raising Agent	Example of Use
.... [1]	.... [1]
.... [1]	.... [1]

3 Name the gas produced by chemical raising agents.

.... [1]

4 This question is about the function of ingredients in choux pastry.

a) When making choux paste, state **two** ingredients that help the pastry rise and puff.

....　　..　　[2]

b) Explain how these ingredients work during baking.

....

....

.... [3]

c) Why is it important to fully cook small choux buns, e.g. profiteroles?

....

....

.... [3]

5 Tick (✓) the correct answer. Raising agents can be classified as:

a) biological, microbial and physical. ☐　　**b)** chemical, enzymic and biological. ☐

c) physical, globular and pathogenic. ☐　　**d)** chemical, physical and biological. ☐　　[1]

Total Marks .... / 15

1 Identify the conditions that food spoilage organisms need to grow.

_____

_____

_____

_____ [3]

2 What are the different signs of food spoilage?

_____

_____

_____

_____ [3]

3 Suggest **two** ways to store dry food such as flour in order to prevent it from spoiling.

_____

_____

_____

_____ [2]

4 Circle the correct option in each of the following sentences.

a) Moulds grow into a(n) **visible** / **invisible** plant. [1]

b) Moulds like **alkali** / **acid** conditions. [1]

c) Moulds are destroyed at temperatures above **50°C** / **70°C**. [1]

d) Moulds **can** / **can't** survive in the refrigerator. [1]

e) Moulds **can** / **can't** survive in a freezer. [1]

5 Tick (✓) **one** answer. Bacteria grow best at a pH level of: [1]

a) between 1.6 and 4.5 ☐

b) between 6.6 and 7.5 ☐

c) between 3.5 and 8.5 ☐

d) between 8 and 9 ☐ [1]

Total Marks _____ / 15

Microorganisms in Food Production

1 Choose the correct words from the options given below to complete the text that follows.

| probiotic | harmful | digestion | single | rapidly |

| cheese | food poisoning |

Bacteria are _____ – celled organisms that are able to reproduce _____. Some are _____ and cause _____ or even death. Some bacteria are harmless and are used in _____ making. _____ bacteria help _____. [7]

2 The process of fermentation is used in the production of which food product? Tick (✓) **one** answer.

a) Bread ☐   **b)** Cheese ☐

c) Yogurt ☐   **d)** Biscuits ☐   [1]

3 Tick (✓) **one** answer. When producing blue cheese, it is treated with:

a) a starter culture. ☐

b) bacteria. ☐

c) mould. ☐

d) brine. ☐   [1]

4 Explain how yeast makes bread rise.

_____

_____

_____

_____

_____

_____

_____

[3]

Total Marks _____ / 12

Bacterial Contamination

1 Complete this table relating to food poisoning.

Name of Bacteria	One Food Source	One Way to Prevent Food Poisoning
a) Salmonella	[1]	[1]
b) Campylobacter	[1]	[1]
c) Bacillus Cereus	[1]	[1]

2 Name **one** symptom of food poisoning.

[1]

3 State **two** conditions that bacteria need in order to reproduce.

[2]

Total Marks _____ / 9

Buying and Storing Food

1 Sally's mother often shops for food during her lunch break.

Describe how she can ensure her food shopping remains at a safe temperature and in good condition until she gets home.

[6]

2 Explain the food hygiene rules that should be followed when storing, preparing and cooking meat.

[6]

Total Marks .... / 12

Preparing and Cooking Food

1 Explain how to reduce the risk of food poisoning when preparing, cooking and storing food in the home.

[6]

2 State **four** food hygiene rules to be followed when preparing and cooking high-risk foods.

[4]

3 List the hygiene and safety rules you would follow when preparing and cooking food.

[8]

Total Marks _____ / 18

Food Choices

1 Tick (✓) the boxes below to show if each statement is **True** or **False**.

Statement	True	False
Vegetarians eat fish.		
Buddhists eat pork.		
Sikhs don't eat beef.		

[3]

2 Which part of the body produces insulin? Tick (✓) **one** answer.

a) Stomach ☐   **b)** Liver ☐   **c)** Pancreas ☐   **d)** Spleen ☐   [1]

3 Coeliac disease is caused by the body's immune system reacting to:

a) sucrose. ☐   **b)** thiamin. ☐   **c)** gluten. ☐   **d)** collagen. ☐   [1]

4 State **three** reasons why a person may be a vegetarian.

....

....

....

[3]

5 Explain the difference between a lacto-vegetarian and a vegan in terms of the foods that each does not eat. Explain why each diet differs.

a) Lacto-vegetarian diet

....

....

....

....

[3]

b) Vegan diet

....

....

....

....

[3]

Total Marks .... / 14

1 Tick (✓) **one** answer. 'Cuisine' relates to:

a) the way in which food is cooked in a kitchen. ☐

b) the range of dishes and foods of a particular country or region. ☐

c) the information listed on a food label. ☐ [1]

2 Which of the following is the name of a British cheese? Circle **one** answer.

Gouda　　　　　　　　　Wensleydale　　　　　　　　　Brie [1]

3 Traditional dishes and foods are important in any society as they originate from the foods grown in that country or region, the local climate and local traditions.

Write the foods in the boxes below next to the correct countries in the table.

| tortilla | Cornish pasty | dhal | minestrone | bouillabaisse | hot pot |

| focaccia | Peking duck | paella | Quiche Lorraine | chow mein | pakora |

Country	Dishes	
a) England		[2]
b) France		[2]
c) Spain		[2]
d) China		[2]
e) India		[2]
f) Italy		[2]

4 Select the cuisine of a country you have studied and complete the table below.

Country Name	Cuisine		
Three main meal dishes	[1]	[1]	[1]
Three vegetables found in recipes	[1]	[1]	[1]

Total Marks _____ / 20

1 Which of the following words is another name for aroma? Circle the correct answer.

feel   smell   touch   taste   [1]

2 How many samples of food would be needed to carry out a 'paired preference test'?
Tick (✓) **one** answer.

a) Two ☐

b) Three ☐

c) Four ☐

d) Five ☐   [1]

3 a) Describe how to set up a tasting area to trial your practical work in the food room.

....

....

....

....

....

....

....

....   [6]

b) What do you understand by a triangle test?

....

....

....   [4]

c) Give **one** example of how you could carry out triangle testing when developing healthier recipes using minced beef.

Name of product: ....   [1]

Samples to trial: ....

....   [2]

Total Marks .... / 15

1 Which food label indicates that the food should no longer be offered for sale in a shop? Tick (✓) **one** answer.

a) Use by ☐   b) Display until ☐

c) Sell by ☐   d) Best by ☐   [1]

2 Which of the following foods is **not** classed as an allergen? Tick (✓) **one** answer.

a) Mustard ☐   b) Pumpkin seeds ☐

c) Celery ☐   d) Peanuts ☐   [1]

3 a) What does GDA stand for?

.... [1]

b) How is GDA displayed on the front of a product label?

....

....

.... [3]

c) Legally, the name of the product must be printed on food packaging. Explain six other items of information that also must be given by law.

....

....

....

....

....

.... [6]

4 Which of the following items does **not** need to be included on a nutrition label on pre-packed foods? Tick (✓) **one** answer.

a) Protein ☐   b) Carbohydrate ☐

c) Salt ☐   d) Sugars ☐   [1]

Total Marks .... / 13

1 In which of these months are British strawberries at their best? Tick (✓) **one** answer.

a) May/June ☐ b) June/July ☐

c) July/August ☐ d) August/September ☐ [1]

2 Tick (✓) **one** answer. The best strategy to eat a healthy and varied diet is to:

a) buy what is on offer. ☐

b) make a shopping list. ☐

c) plan a shopping diary, make a shopping list, and keep to it. ☐

d) eat lots of fruit and vegetables. ☐ [1]

3 a) Explain the benefits for young people of being active for at least 60 minutes every day.

[4]

b) To maintain a healthy weight, what eating model does the UK Government suggest that people use?

[1]

c) What factors will make a meal more enjoyable to eat?

[3]

d) Explain **four** factors that affect the eating patterns within a household.

[4]

Total Marks _____ / 14

1 Choose the correct words from the boxes below to complete the following sentences.

| transport | process | dispose | energy | carbon footprint | carbon dioxide |

At each stage of a product's lifecycle _____ is needed to

_____ , _____ and _____ of the product;

_____ is produced as a byproduct of energy use. The _____ is

the calculation of the carbon dioxide produced throughout a product's life.   [6]

2 Choose the correct words from the boxes below to complete the following sentences.

| livestock | soil erosion | pasture | wasteland | plantations | deforestation |

_____ occurs when trees are cut down. Cleared land is used as

_____ for _____ and _____ of commodities

and settlements. Deforested regions typically suffer _____ and frequently

degrade into _____ .   [6]

3 When you throw away food, you waste not only the food but also the resources, such as energy, fuel and water, that went into growing, harvesting, transporting and storing the food. Discarded food then goes on to produce methane in landfill sites. Explain what every person can do to prevent this food waste.

_____

_____

_____

_____

_____

_____

_____

     [6]

4 Name **two** dishes that make use of leftover food.

_____   [2]

Total Marks _____ / 20

Food Provenance and Production Methods

1 Why is traceability important?

[4]

2 Explain the differences between the hens that lay battery, free range, barn and organic eggs.

Battery

[2]

Free-range

[2]

Barn

[2]

Organic

[2]

3 What concerns are there about GM (Genetically Modified) food production? Tick (✓) the correct answers.

a) It is more expensive.

b) There is a possibility of new strains of microorganisms developing.

c) It is altering and playing with nature.

d) It is less resistant to plant disease.

e) It is not monitored.

[2]

Total Marks .... / 14

1 If large areas of rainforest are cut down, which gas will build up?

.... [1]

2 Write about **three** different things that can be done to tackle the sustainability of a food source.

....

....

.... [3]

3

FAIRTRADE

a) What does this logo mean?

....

.... [2]

b) Give **two** examples of foods that could display this logo.

.... [1]

4 What are the advantages of buying a pack of chicken that displays the Red Tractor logo?

....

....

....

....

....

....

.... [8]

5 Beef burgers are a very popular takeaway food and use beef, which means that more cattle are reared to supply the fast food industry. Explain what effect this demand for beef supply is having on our climate.

....

....

....

.... [2]

Total Marks .... / 17

Food Production

1 What is homogenised milk?

....

.... [2]

2 What can happen if cream is heated?

.... [1]

3 a) Name **three** regional varieties of cheese from the UK.

.... [3]

b) Name **three** French cheeses.

.... [3]

4 Name **six** different types of flour made from wheat.

....

.... [6]

5 Choose the correct words from the boxes to complete the following sentences.

| gluten | sugar | yeast | proving | kneading | glaze | dough | liquid |

Bread is made by mixing strong flour (which is high in ....) with

.... and a raising agent such as .... The yeast ferments

with .... and warm water, and then when added to the flour and salt it makes

a .... The dough is then worked by a process called ....

The dough is then allowed to rise further by standing covered in a warm environment.

This is known as .... The bread dough is then shaped and finished with a

.... before baking. [8]

6 What type of flour is used to make pasta? What is it made from?

.... [2]

Total Marks .... / 25

1 Name the **four** main types of food preservation.

....

.... [4]

2 You have been growing your own fruit and have a large crop of apples. How can you preserve them? Explain your method, and the shelf life of your chosen method.

Method ....

Shelf life ....

.... [2]

3 What is the main advantage of low-temperature storage when considering the nutritional value of food?

.... [1]

4 Often meats or fish are 'smoked'. What does this mean?

....

.... [2]

5 Years ago, before electricity, there were no fridges and freezers, so how did people preserve the produce from gardens to last through the seasons?

....

....

....

.... [4]

6 Roll mops are a traditional fish dish. What method of preservation do they use?

.... [1]

7 How is bacon preserved?

.... [2]

Total Marks .... / 16

Collins

GCSE
Food Preparation And Nutrition

Practice Paper 1

Materials

Time allowed: 1 hour 45 minutes

For this paper you must have:

- a black pen
- a pencil.

Instructions

- Use black ink or black ball-point pen.
- Answer **all** questions.
- You must answer the questions in the spaces provided. Do not write outside the box around each page or on blank pages.
- Do all rough work in this answer book. Cross through any work you do not want to be marked.

Information

- The marks for questions are shown in brackets.
- The maximum mark for this paper is 100.
- You are reminded of the need for good English and clear presentation in your answers.

Name: ....

Practice Exam Paper 1

Section A consists of multiple choice questions.

There are 20 marks available.
Answer all questions.

For each question you should shade in **one** box.

An example is shown below.

Which food is high in fat?

A Bread ☐

B Cheese ■

C Broccoli ☐

D Apple ☐

Question 1 is about diet, nutrition and health.

0 1 . 1 Fats are made up of which **two** components?

A Cholesterol and acetic acid ☐

B Fatty acids and glycerol ☐

C Glycerol and glucose ☐

D Butter and margarine ☐   [1 mark]

0 1 . 2 Plants manufacture carbohydrate by a process known as:

A osmosis. ☐

B osteoporosis. ☐

C combustion. ☐

D photosynthesis. ☐   [1 mark]

`0 1`·`3` A deficiency of iron in the diet can cause anaemia. Which of the following foods is a dietary source of iron?

- **A** Olive oil ☐
- **B** Soda bread ☐
- **C** Liver ☐
- **D** Cottage cheese ☐ [1 mark]

`0 1`·`4` Coronary Heart Disease is caused by a build-up of what substance in the arteries?

- **A** Cholesterol ☐
- **B** Bile ☐
- **C** Riboflavin ☐
- **D** Glycerol ☐ [1 mark]

Question 2 is about food safety.

`0 2`·`1` After which **one** of the following activities is it most important for the food handler to wash their hands?

- **A** Washing lettuce ☐
- **B** Spreading butter on bread ☐
- **C** Peeling fruit ☐
- **D** Preparing raw chicken ☐ [1 mark]

`0 2`·`2` Which of the following is most likely to cause cross contamination?

- **A** Using ready-to-eat foods within their use-by date ☐
- **B** Placing ready-to-eat foods above raw foods in a fridge ☐
- **C** Using the same knife to cut raw chicken and cooked ham ☐
- **D** Storing raw chicken in a covered container at the bottom of the fridge ☐ [1 mark]

Practice Exam Paper 1

0 2 · 3 What is the temperature regarded as the danger zone?

A   0–5 °C ☐

B   0–63 °C ☐

C   5°C–63 °C ☐

D   5–50 °C ☐ **[1 mark]**

0 2 · 4 Which microorganism is used in the manufacture of bread?

A   Yeast ☐

B   Flour ☐

C   Salt ☐

D   Salmonella ☐ **[1 mark]**

Question 3 is about food science.

0 3 · 1 Which **one** of the following is a true statement?

Choux paste is raised by:

A   flour as it gelatinises during cooking. ☐

B   water as it turns to steam during baking. ☐

C   beating during the adding of the flour. ☐

D   eggs enriching the choux paste. ☐ **[1 mark]**

0 3 · 2 Which type of flour has the highest chemical raising agent content?

A   Self-raising flour ☐

B   Soya flour ☐

C   Maize flour ☐

D   Plain wheat flour ☐ **[1 mark]**

0 3 · 3 Complete this sentence.

The main function of egg in pancake batter is:

A to gelatinise during cooking. ☐

B to bind the batter mixture. ☐

C to set the mixture by coagulating. ☐

D to brown the mixture by caramelising. ☐ [1 mark]

0 3 · 4 When making whisked sponge, which gas is trapped during whisking?

A Lecithin ☐

B Sulphur ☐

C Air ☐

D Carbon dioxide ☐ [1 mark]

Question 4 is about food provenance.

0 4 · 1 What is intensive farming?

A Farmers who work 24 hours a day, seven days a week. ☐

B Small farms that specialise in rearing just one animal or crop. ☐

C Large-scale farms that focus on large-scale production of crops

or animals in a short period of time. ☐

D Farms that have a large variety of animals or crops. ☐ [1 mark]

0 4 · 2 Deforestation contributes to global warming with excesses of which gas?

A CO_2 ☐

B Methane ☐

C Carbon monoxide ☐

D Oxygen ☐ [1 mark]

0 4 . 3 What is hydroponics?

A   An irrigation system for fields

B   Nutrient-rich liquids in which to grow plants

C   Ponds on farms that provide animal drinking water

D   Water drip feeders for hens in barns   [1 mark]

0 4 . 4 When making bread, which type of flour do we need to use?

A   Durum wheat 00 flour

B   Plain flour

C   Strong flour

D   Self-raising flour   [1 mark]

Question 5 is about food choices.

0 5 . 1 If you have Coeliac disease, which **one** of the following foods are you **not** able to eat?

A   Rice

B   Potatoes

C   Bread

D   Cabbage   [1 mark]

0 5 . 2 Which **one** of the following foods does a lacto-vegetarian **include** in their diet?

A   Milk

B   Eggs

C   Fish

D   Meat   [1 mark]

0 5 · 3 Which **one** of these is a **true** definition of a shopping list?

    A List of offers on sale ☐

    B List of foods to buy ☐

    C List of dishes to make ☐

    D List of foods people like ☐ **[1 mark]**

0 5 · 4 Which term describes how food **feels** in the mouth when eating?

    A Aroma ☐

    B Taste ☐

    C Texture ☐

    D Appearance ☐ **[1 mark]**

Section B

Answer all questions in this section.
There are 80 marks available.

Question 6 is about diet, nutrition and health.

`0 6` · `1` One of the most common medical procedures for primary school children is dental extraction. Dentists are increasingly concerned that this problem is growing. Explain in detail how sugar can cause tooth decay.

[4 marks]

`0 6` · `2` Suggest four ways in which children's sugar consumption can be reduced.

....

....

....

....

[4 marks]

0 6 · 3 Name **two** other diseases that can be caused by a high intake of sugar. Discuss how too much sugar causes these diseases.

....

....

....

....

....

....

....

....

[6 marks]

0 6 · 4 Sugar is a carbohydrate. Name the **two** other carbohydrate groups and give **two** examples of foods from each.

....

....

....

[6 marks]

0 6 · 5 Consider and explain the nutritional requirements of primary school children, detailing each nutrient group.

....

....

....

....

....

....

....

....

[12 marks]

Question 7 is about cooking food.

The information below shows a recipe for béchamel sauce.

400 ml milk	25 g piece of carrot	1 stalk of celery	6 black peppercorns
1 bay leaf	Seasoning	25 g margarine	25 g flour

Using the information above, answer the following questions.

0 7 . 1 Which herbs and spices are used in this recipe? Give **one** example of each.

Herb ....

Spice ....

[2 marks]

0 7 . 2 Explain why béchamel sauce is not suitable for someone who is lactose intolerant.

....

....

[2 marks]

0 7 . 3 Explain how heat is transferred when the vegetables and peppercorns are infused in the milk.

....

....

....

[3 marks]

0 7 . 4 Explain how the liquid is thickened by the roux.

....

....

....

[3 marks]

0 7 · 5 The table below shows **two** dishes that use flour as an ingredient. For each dish give **one** function of the flour and **one** description of the function.

Name of Dish	Function	Description
Choux pastry		
Bread		

[6 marks]

Question 8 is about food provenance.

0 8 · 1 Fresh milk is processed into several different types of milk. Name **four** different varieties of milk and explain how they are different from each other, e.g. in homogenised milk the fat is broken up and dispersed through the milk so it doesn't reform as a layer.

[8 marks]

0 8 · 2 Secondary processing of milk produces a whole variety of milk-based products. Name **two** milk-based products and describe how they are made.

[6 marks]

Question 9 is about understanding recipes.

Information about two pasta sauces is given below.

Information about nutrients

Ingredients in quantity order	Ingredients	energy kcal	protein g	carbohydrate g	unsaturated fat g	saturated fat g	dietary fibre g	sugar g	salt g
Sauce A Tomato sauce with black olives, aubergine and ricotta	tomatoes aubergine olive oil black olives ricotta onion oregano	148 kcal	3.2 g	7 g	9.8 g	2.2 g	4 g	3.2 g	1.9 g
Sauce B Chorizo and chilli sauce	tomatoes onion red pepper chorizo olive oil smoked paprika black pepper lemon juice	66 kcal	1.7 g	7.2 g	3 g	0.7 g	1 g	4.8 g	0.5 g

0 9 · 1 With reference to the ingredients and nutrient content of each of the sauces, evaluate the suitability of these sauces for obese people.

Give reasons for your choices.

....

....

....

....

....

[8 marks]

Question 10 is about food preparation and food safety.

1 0 . 1 The table below shows some problems seen when food is prepared. Complete the table to show **two** different causes of each problem.

Problem	Causes
Victoria sandwich not rising	Cause 1
	Cause 2
Lumpy cheese sauce	Cause 1
	Cause 2

[4 marks]

1 0 . 2 'High-risk foods are most likely to cause food poisoning.' Explain this statement.

[6 marks]

END OF QUESTIONS

Collins

GCSE
Food Preparation And Nutrition

Practice Paper 2

Materials

Time allowed: 1 hour 45 minutes

For this paper you must have:

- a black pen
- a pencil.

Instructions

- Use black ink or black ball-point pen.
- Answer **all** questions.
- You must answer the questions in the spaces provided. Do not write outside the box around each page or on blank pages.
- Do all rough work in this answer book. Cross through any work you do not want to be marked.

Information

- The marks for questions are shown in brackets.
- The maximum mark for this paper is 100.
- You are reminded of the need for good English and clear presentation in your answers.

Name: ....

Section A consists of multiple choice questions.

There are 20 marks available.
Answer all questions.

For each question you should shade in **one** box.

An example is shown below.

Which food is high in fat?

A Bread [ ]

B Cheese [■]

C Broccoli [ ]

D Apple [ ]

Question 1 is about diet, nutrition and health.

`0 1 . 1` A person who is unable to eat wheat products is:

A vegan. [ ]

B coeliac. [ ]

C diabetic. [ ]

D anaemic. [ ] **[1 mark]**

`0 1 . 2` Citrus fruits are a good source of:

A iron. [ ]

B calcium. [ ]

C vitamin D. [ ]

D vitamin C. [ ] **[1 mark]**

Practice Exam Paper 2

0 1 · 3 Which of the following is not a function of vitamins?

A   Prevent illness and maintain good health

B   Protect the body's internal organs

C   Aid building and repair in the body

D   Control the release of energy by the body   **[1 mark]**

0 1 · 4 Which of the following contains a vegetable source of fat?

A   Rapeseed oil

B   Cream

C   Butter

D   Lard   **[1 mark]**

Question 2 is about food safety.

0 2 · 1 Which food poisoning bacteria are commonly found on human skin?

A   Clostridium botulinum

B   Salmonella enteritidis

C   Staphylococcus aureus

D   Clostridium perfringens   **[1 mark]**

0 2 · 2 Which of the following does bacteria need in order to multiply?

A   Light

B   Moisture

C   Sugar

D   Salt   **[1 mark]**

0 2 · 3 What is the required temperature for fridge storage?

    A 1–4 °C (below 5 °C) ☐

    B 5–15 °C (below 15 °C) ☐

    C 0–63 °C ☐

    D Minus 15 °C ☐ **[1 mark]**

0 2 · 4 Why are use-by dates put on high-risk foods?

    A This is the date that the food will be at its best to eat. ☐

    B The food is perishable and may be unsafe to eat after that date. ☐

    C So that the food is not eaten by high-risk groups. ☐

    D So that you know when to put the item in the freezer. ☐ **[1 mark]**

Question 3 is about food science.

0 3 · 1 Which **one** of the following is a true statement?
Food is cooked to:

    A make it safe to eat and give variety in the diet. ☐

    B incorporate carbohydrates, fats and proteins. ☐

    C soften food and prevent lumps from forming. ☐

    D make it suitable to eat immediately after cooking. ☐ **[1 mark]**

0 3 · 2 Which type of cooking method uses the most water?

    A Stir frying ☐

    B Braising ☐

    C Steaming ☐

    D Boiling ☐ **[1 mark]**

$0\ 3\ \cdot\ 3$ Complete this sentence.

The main function of grilling sliced bread is:

A caramelisation of the surface. ☐

B dextrinisation of the starch. ☐

C coagulation of the gluten. ☐

D gelatinisation of the starch. ☐ **[1 mark]**

$0\ 3\ \cdot\ 4$ When cooking in an oven, which main heat transfer is used?

A Radiation ☐

B Microwave ☐

C Conduction ☐

D Convection ☐ **[1 mark]**

Question 4 is about food provenance.

$0\ 4\ \cdot\ 1$ 4.1 What does MAP stand for? ☐

A Made Abroad Product ☐

B Modified Atmospheric Packaging ☐

C Microwavable Appetising Product ☐

D Modified Additive Product ☐ **[1 mark]**

$0\ 4\ \cdot\ 2$ Often when purchasing fish, the packaging says 'from sustainable sources'. What does this mean?

A The fish have been caught whilst young so they are more tender. ☐

B The holes in the fishing nets are small to ensure that no fish get away. ☐

C Fish are caught by fisherman observing fish quotas. ☐

D The fish have been processed and chilled on the ship after being caught. ☐ **[1 mark]**

`0 4`·`3` Which of the following does not contribute to global warming?

    A Most households owning and using refrigerators ☐

    B Recycling of packaging ☐

    C Food production and transporting products ☐

    D Home-grown fruit and vegetables ☐ **[1 mark]**

`0 4`·`4` One of the effects of climate change can be flooding. Which of the following is not a consequence of this?

    A Livestock drowned ☐

    B Land polluted by sewage and debris ☐

    C Pests invading crops and destroying them ☐

    D Soil and nutrients are washed away ☐ **[1 mark]**

Question 5 is about food choices.

`0 5`·`1` Which sensory test finds out how much someone likes the taste of a food?

    A Triangle ☐

    B Ranking ☐

    C Rating ☐

    D Paired preference ☐ **[1 mark]**

`0 5`·`2` Which food is traditionally eaten at Jewish Passover?

    A Unleavened bread ☐

    B Baklava ☐

    C Hot cross buns ☐

    D Pumpkin pie ☐ **[1 mark]**

Practice Exam Paper 2

0 5 · 3 Which of the following foods is **not** one of the 14 allergen foods?

    A Soybeans ☐

    B Crustaceans ☐

    C Beetroot ☐

    D Cow's milk ☐ **[1 mark]**

0 5 · 4 Which of these ingredients is likely to be the **least** expensive when shopping for food on a budget?

    A Fish ☐

    B Strawberries ☐

    C Asparagus ☐

    D Potatoes ☐ **[1 mark]**

Section B

Answer all questions in this section.
There are 80 marks available.

Question 6 is about diet, nutrition and health.

| 0 | 6 | · | 1 | Diets which are high in saturated fats are directly linked to several serious health problems.

> Sam is a teenager. He has a fried breakfast every day: two economy sausages, streaky bacon rashers, fried egg and fried wholemeal bread with a hot chocolate drink made with whole milk.

Explain how the macronutrient content of the breakfast provides Sam with energy.

[6 marks]

0 6 . 2 In the UK diets high in saturated fats are directly linked to several serious health problems.

Assess the various factors that contribute to high fat intake and explain how high fat diets in childhood and teenage years may put future health at risk.

_____

_____

_____

_____

_____

_____

_____

_____

_____

_____

_____

_____

_____

[12 marks]

0 6 . 3 Describe and explain **three** functions of fat in the body.

_____

_____

[6 marks]

0 6 . 4 Fat and oils have different chemical compositions.

Describe the different make-up of the following types of fat, giving **two** examples of each type.

Saturated

Unsaturated

[8 marks]

Question 7 is about cooking food.

The information below shows a recipe for shortbread.

150 g plain four	100 g butter	50 g caster sugar
Grated lemon zest	Ground cinnamon	

Using the information above, answer the following questions.

0 7 . 1 Which flavouring and which spice is used in the recipe?

Flavouring ....

Spice ....

[2 marks]

0 7 . 2 Explain why this recipe is not suitable for someone who needs to reduce saturates in their diet.

....

....

[2 marks]

0 7 . 3 Explain how heat is transferred during cooking the shortbread.

....

....

[3 marks]

0 7 · 4 Explain the changes that occur during baking the shortbread.

[3 marks]

0 7 · 5 The table below shows **two** dishes that use margarine as an ingredient. For each dish give **one** function of the margarine and **one** description of the function.

Name of Dish	Function	Description
Pastry tarts		
Sponge cake		

[6 marks]

Question 8 is about food provenance.

0 8 · 1 Farmers' markets are very popular these days. Discuss the advantages of buying from farmers' markets.

[6 marks]

0 8 . 2 Many farmers' markets sell fresh farm eggs, but these can be laid by hens in a variety of housing conditions. Describe **three** different ways of egg farming.

[6 marks]

0 8 . 3 How you can ensure you are buying a quality product when buying fresh eggs?

[2 marks]

Question 9 is about understanding recipes.

Information about two gluten-free sausages is given below.

Nutrients per 100 g

Ingredients in quantity order	Ingredients	energy kcal	protein g	carbohydrate g	unsaturated fat g	saturated fat g	dietary fibre g	sugar g	salt g
Sausage A Quorn Gluten-free	Quorn onion rice flour potato starch egg white black pepper thyme sage	197 kcal	15.7 g	10.3 g	8.3 g	0.8 g	5.5 g	1 g	1.2 g
Sausage B Musks Gluten-free	pork rice salt spices	210 kcal	15.0 g	3.5 g	9.2 g	5.5 g	0.7 g	0.7 g	1.8 g

0 9 . 1 With reference to the ingredients and nutrient content of each of the sausages, evaluate the suitability of these sausages for a person who is coeliac.

Give reasons for your choice.

....

....

....

....

....

....

....

....

[8 marks]

Question 10 is about food preparation and food safety.

| 1 | 0 | · | 1 | The table below shows some problems seen when food is prepared.

Complete the table to show **two** different causes of each problem.

Problem	Causes
Bread not rising	Cause 1
	Cause 2
Quiche not setting	Cause 1
	Cause 2

[4 marks]

1 0 . 2 Explain how you can prevent food poisoning when storing food in the home.

....

....

....

....

....

....

....

....

....

[6 marks]

END OF QUESTIONS

Answers

Page 148: Knife Skills

1. a) **[1]**
2. b) **[1]**
3. Name of knife hold 1: Bridge hold **[1]**
   Explanation: Form a bridge with thumb and index finger, hold item flat side down on chopping board, position knife under the bridge and cut firmly downwards. **[1]**
   Name of knife hold 2: Claw grip **[1]**
   Explanation: Place item to be cut flat side down on chopping board, shape hand into a claw, tuck thumb inside fingers, rest the claw on item to be sliced, use other hand to slice the item, moving clawed fingers away as cutting progresses. **[1]**
4. **Any four from**: Carry knife pointing downwards **[1]**; Handle should be grease-free **[1]**; Don't put in washing-up bowl **[1]**; Keep clean **[1]**; Keep sharp **[1]**; Do not leave on edge of surface **[1]**; Use correct knife for the job to be done **[1]**.

Page 149: Fish

1. c) **[1]**
2. a) **[1]**
3. d) **[1]**
4. Enrobing **[1]**; Coating **[1]**
5. Fish cooks quickly because the muscle **[1]** is short and the connective tissue **[1]** is thin.
   The connective tissue is made up of collagen **[1]** and will change into gelatine **[1]** and coagulate **[1]** at 60 °C **[1]**.
6. Classification of fish: round
   One example of fish: **any one of**: Cod **[1]**; Haddock **[1]**; Whiting **[1]**; Pollock **[1]**; Coley **[1]**.
   Chef is cutting into the top of the fish on one side of the tail to detach the backbone from the head to the tail **[1]**; Chef has left the head on the fish **[1]**.

Page 150: Meat

1. d) **[1]**
2. c) **[1]**
3. Meat is a muscle **[1]** composed of cells consisting of fibres **[1]**, held together by connective tissue **[1]**.
4. **Any one of**: Leg **[1]**; Shoulder **[1]**.
5. **Any two from**: Stewing **[1]**; Braising **[1]**; Casserole **[1]**; Pot-roasting **[1]**.
6. The browning **[1]** of meat is caused by a reaction with natural sugars **[1]** and proteins to produce a dark colour. This occurrence is called the Maillard **[1]** reaction or non-enzymic browning. As the meat cooks the proteins coagulate **[1]** and produce a firm texture. Collagen is broken down into gelatine **[1]**.

Page 151: Prepare, Combine and Shape

1. c) **[1]**
2. a) **[1]**
3. Whisking involves adding air into the mixture while mixing **[1]**; Stirring does not involve adding air into the mixture while mixing **[1]**.
4. **Any two from**: Consistency of depth **[1]**; Consistency of size **[1]**; Consistency of shape **[1]**.
5. The following mixing skills should be used to ensure the smooth consistency of the batter: Cream **[1]** together the fat and sugar **[1]**; Beat **[1]** the eggs; Fold **[1]** in the flour. The following shaping skills should be used to ensure quality of finish: Use cake tins **[1]** to mould **[1]** the shape of the cake; Pipe **[1]** the cream for a good decorative finish **[1]**.
6. Use a burger mould (or burger press) **[1]**; Weigh the mixture to make sure the same quantity is used for each burger **[1]**

Page 152: Dough

1. c) **[1]**
2. b) **[1]**
3. Rubbing in **[1]**
4. To bind the dough together **[1]**
5. Weigh all the ingredients accurately **[1]**.
   Place fat **[1]** and water **[1]** in a pan and bring to a rolling boil **[1]**. Add sieved flour **[1]** immediately and mix well to form a paste **[1]**. Cool **[1]** then add beaten eggs gradually to a heavy dropping **[1]** consistency. Cook in a hot oven.
6. Glutenin **[1]**; Gliadin **[1]**

Page 153: Protein and Fat

1. growth **[1]**; repair **[1]**
2. Possible answers, **any two from**: Cheese **[1]**; Milk **[1]**; Eggs **[1]**; Pulse vegetables **[1]**; Soya **[1]**; TVP **[1]**; Mycoprotein (Quorn) **[1]**; Nuts **[1]**
3. Possible answers, **any three from**: Meat **[1]**; Fish **[1]**; Cheese **[1]**; Eggs **[1]**; Milk **[1]**; Soya **[1]**.
4. Lentil soup and bread **[1]**
5. c) **[1]**
6. Possible answers, **any one of**: Any type of oil, e.g. groundnut oil **[1]**; Vegetable oil **[1]**; Olive oil **[1]**; Sunflower oil **[1]**; Any solid fat that has been melted **[1]**.
7. Provides concentrated energy. **[1]**; Provides body with vitamins A, D, E and K. **[1]**
8. Fatty Acids **[1]**; Glycerol **[1]**
9. Hydrogenation is the name given to the process that makes solid fat from a liquid oil **[1]**.

Page 154: Carbohydrate

1. Sugars **[1]**; Starches **[1]**; Non-Starch Polysaccharide (dietary fibre) **[1]**
2. Possible answers, **any one of**: Glucose **[1]**; Galactose **[1]**; Fructose **[1]**.
3. digested **[1]**; energy **[1]**
4. a) **Any one of**: Wholemeal bread **[1]**; Granary bread **[1]**.
   b) **Any one of**: Branflakes **[1]**; All Bran **[1]**; Fruit and fibre, etc. **[1]**.
   c) Jacket potatoes **[1]**
5. Dental decay **[1]**
6. **Any two from:** The body will start to use protein and fat as an energy source **[1]**; Weight loss **[1]**; Lack of energy **[1]**; Poor digestive health **[1]**.
7. Starches **[1]** have to be digested into sugars **[1]** before absorption **[1]** – this is slow energy release **[1]**.
8. Constipation **[1]**

Page 155: Vitamins

1. Possible answers, **any two from**: Liver **[1]**; Whole milk **[1]**; Cheese **[1]**; Green leafy vegetables **[1]**; Carrots **[1]**.
2. The water soluble group of vitamins **[1]**
3. Weak bones in children, that bend under body weight **[1]**; Associated with a vitamin D deficiency **[1]**
4. Skin cracking around the mouth **[1]**
5. Antioxidants protect us from pollutants in the environment **[1]**.
6. A type of vitamin D **[1]** formed by action of sunlight on the skin **[1]**
7. When making a salad, to avoid the loss of vitamin C (ascorbic acid **[1]**) through oxidation **[1]**, prepare just before serving **[1]** and avoid excess cutting **[1]**.
8. b) **[1]**

Page 156: Minerals and Water

1. d) **[1]**
2. **Any one of**: Strong bones **[1]**; Strong teeth **[1]**; To enable clotting of blood **[1]**; For nerves and muscles **[1]**; Works with Vitamin D **[1]**; Prevents rickets/brittle bones/osteoporosis **[1]**.
3. **Any one of**: High blood pressure **[1]**; Heart disease **[1]**; Strokes **[1]**.
4. Flavour **[1]**
5. A lack of iron in the diet can cause iron deficiency anaemia **[1]** and symptoms include tiredness **[1]**.
6. Between six and eight glasses **[1]**
7. Fluoride is important for strengthening teeth against decay **[1]**
8. a) **[1]**

Page 157: Making Informed Choices

1. Fruit and vegetables **[1]**; Potatoes, bread, rice, pasta and other starchy carbohydrates **[1]**; Beans, pulses, fish, eggs, meat and other proteins **[1]**;

Dairy (and alternatives) [1]; Oils and Spreads [1]

2. Human milk [1] provides babies with all their nutritional requirements, except for iron [1]. Babies are born with a supply of this stored in their liver [1].
3. A mother's first milk is called colostrum and it is full of antibodies [1].
4. For the development of the neural tube of the foetus [1]. This can prevent the condition spina bifida [1].
5. c) [1]
6. The Eatwell Guide [1] shows the proportions of food groups that should be eaten daily for a well-balanced [1] diet.
7. Eat plenty of fibre-rich foods [1]; for example, **any one of**: wholegrain cereals/wholemeal bread/wholegrain breakfast cereals/wholemeal pasta/wholemeal flour/fruit/ vegetables/ dried fruit/nuts/seeds/ beans, peas/lentils [1].

Page 158: Diet, Nutrition and Health

1. Coronary Heart Disease [1] is linked to diets high in saturated fats [1], which make cholesterol [1] in the blood.
2. c) [1]
3. Five to seven portions (allow 5 portions) [1]
4. **Any three from**: Weight gain/obesity [1]; Can produce high/bad cholesterol [1]; Can block arteries [1]; Angina [1]; Coronary Heart Disease (CHD)/ heart disease/heart attack [1]; Higher consumption of trans fats/higher risk of cancer [1].
5. In Type 2 diabetes, too little or no insulin [1] is produced resulting in high levels of sugar [1] in the blood [1].
6. b) [1]
7. Age [1]; Gender [1]; Body size [1]

Page 159: Cooking of Food, Heat Transfer and Selecting Appropriate Cooking Methods

1. c) [1]
2. Cooking uses heat [1]; in order to change the texture, flavour and colour of food [1]; and to improve palatability [1].
3. Dry method, change to nutritional content: Reduces fat [1]; Example: Grilling sausages.
   Water-based method, change to nutritional content: (moist method) vitamin C loss [1]; Example: Boiling potatoes/cabbage [1].
4. Cooking methods can alter the sensory [1] properties of food by adding crispness or softening.
   The right cooking methods can conserve [1] vitamins, or add [1] energy value.
   When meat needs cooking, long, slow moist [1] cooking is best for tough [1] cuts of meat.
   If the meat is tender a quick [1] and hot [1] method of cooking such as grilling can be used.

Page 160: Proteins and Enzymic Browning

1. c) [1]
2. a) A marinade is a liquid [1]; made from flavoursome and acidic ingredients that is used to soak [1]; foods prior [1] to cooking.
   b) A marinade is used to add flavour to foods from ingredients such as garlic, chillies, herbs and/or spices [1]; A marinade using acidic ingredients such as lemon juice, vinegar or buttermilk is used to make ingredients such as meat or fish more tender [1]; Marinades help to add moistness in foods that otherwise might be dry [1].
   c) Chicken has a fairly bland, mild flavour [1]; therefore a marinade would add flavour [1]; The meat, chicken, would be made tender, kept juicy and not be tough [1]; and would cook quickly [1]
   d) BBQ [1]; Grill [1]
3. a) True [1]
   b) True [1]
   c) False [1]
   d) True [1]

Page 161: Carbohydrates

1. a) [1]
2. b) [1]
3. semi-skimmed milk: liquid for the sauce [1]
   wheat flour: thickener [1]
   margarine: fat for the roux [1]
   seasoning: salt and pepper flavour [1]
   Red Leicester cheese: main protein/ cheese flavour [1]
   macaroni: pasta/carbohydrate [1]
4. 1 Boil the pasta [1]
   2 Drain the pasta [1]
   3 Prepare the sauce [1]
   4 Use the roux method or use the all-in-one method to prepare the sauce [1]
   5 Assemble the pasta and the sauce [1]
   6 Au gratin option [1]
5. Mix the roux (or use the all-in-one method) [1]; Heat thickens by starch gelatinising [2]; and beat – agitation ensures smoothness [2]

Page 162: Fats and Oils

1. b) [1]
2. a) The vinegar [1]; and the melted butter [1] would not mix easily – they would separate on standing
   b) Seasoning makes the sauce taste better [1]; Salt and pepper are commonly used as seasoning [1]
   c) The egg yolk contains lecithin [1]; which emulsifies [1]; the butter and vinegar to create a stable sauce [1]
3. a) binds [1]
   b) dextrinisation [1]
   c) shorten [1]
   d) steam [1]
   e) gluten [1]

Page 163: Raising Agents

1. a) [1]
2. Names of chemical raising agents: Bicarbonate of soda [1]; Baking powder [1]
   Examples of use, **any two of**: Scones [1]; Gingerbread [1]; Biscuits [1].
3. Carbon dioxide [1]
4. a) Water [1]; Eggs [1]
   b) During baking the water turns to steam [1]; The eggs expand [1]; Before setting/coagulating [1], holding the risen shape
   c) Fully cooked buns are dry inside [1]; Steam has escaped [1]; This prevents collapse [1]
5. d) [1]

Page 164: Microorganisms, Enzymes and Food Spoilage

1. **Any three from**: Warm temperature/ 37 °C [1]; Moisture [1]; Food [1]; Time [1]; Neutral pH [1]; May need oxygen [1].
2. **Any three from**: Mould grows [1]; Flavour changes (souring) [1]; Bacterial contamination [1]; Physical contamination from dirty machinery or careless food handlers [1]; Contamination by flies, cockroaches, mice, rats, mites, domestic animals [1]; Contamination by chemicals/radiation/ pollution [1]; Colour changes [1]; Texture changes [1]; Unpleasant odour [1].
3. **Any two from**: Cupboard should be free from vermin and pets [1]; Wash shelves regularly/deal with spills immediately [1]; Make sure storage containers are clean [1]; Do not top up existing stock with new [1]; Store dry foods in airtight containers/ sealed packets [1]; Keep a check on approximate storage times/best before dates [1]; Store in a cool dry place [1].
4. a) visible [1]
   b) acid [1]
   c) 70 °C [1]
   d) can [1]
   e) can't [1]
5. b) [1]

Page 165: Microorganisms in Food Production

1. Bacteria are single [1] -celled organisms that are able to reproduce rapidly [1]. Some are harmful [1] and cause food poisoning [1] or even death. Some bacteria are harmless and are used in cheese [1] making. Probiotic [1] bacteria help digestion [1].
2. a) [1]
3. c) [1]
4. Air bubbles are trapped and distributed throughout the bread dough as it is mixed and kneaded [1]; The yeast absorbs the starches and sugars in the flour, turning them into alcohol and carbon dioxide gas

[1]; The gas inflates the air bubbles, causing the bread to rise [1].

Page 166: Bacterial Contamination

1. a) Food source, **any one of**: Raw meat, poultry/chicken [1]; Eggs [1]; Cooked meat [1]; Dairy foods [1]; Cheese [1]; Mayonnaise [1]; Bean sprouts.
   Prevention, **any one of**: Wash hands after handling raw meat, eggs etc. [1]; Hard boil eggs/avoid lightly cooked or raw eggs [1]; Defrost chicken before cooking [1]; Cook meat, poultry/chicken thoroughly [1]; Boil bean sprouts before use.
   b) Food source, **any one of**: Meat [1]; Shellfish [1]; Untreated water [1]; Washing raw poultry [1].
   Prevention, **any one of**: Take measures to prevent transmission between humans [1]; Raw meat and poultry **must not** be washed, as this spreads the bacteria [1].
   c) Food source, **any one of**: Cooked rice [1]; Herbs and spices [1]; Starchy food products [1].
   Prevention, **any one of**: Do not reheat rice dishes [1]; Cool cooked rice immediately after cooking when making salads etc. [1]; Do not keep herbs and spices past use-by date [1].
2. **Any one of**: Diarrhoea [1]; Dehydration [1]; Headache [1]; High/low temperature [1]; Sickness/vomiting [1]; Stomach ache/cramps/nausea/feel sick [1].
3. **Any two from**: Food/nutrients [1]; Moisture/damp [1]; Oxygen/air [1]; Time [1]; Warmth [1].

Page 167: Buying and Storing Food

1. **Any six from**: Use a cool bag/cool box/polystyrene material to insulate the cold food from room temperature [1]; Keep the food covered in the boot of the car, if transporting home by car [1]; Ensure all refrigerated food is kept together in the same bag [1]; Do not buy frozen foods if you can't get them home quickly as they will defrost [1]; Use a collection service to get refrigerated food home as quickly as possible [1]; Use a home delivery service to get refrigerated food straight from refrigerated transport into your fridge [1]; Take the quickest route home so that refrigerated food is out of the fridge for as short a time as possible [1]; Store refrigerated food in a fridge, if available, at work [1]; Buy food with good packaging to keep it in shape, avoid squashing etc., e.g. eggs [1]; Park car in a shaded cool spot/not sunny area to keep the internal temperature of the car down so that temperature gain in refrigerated food is kept to a minimum [1].

2. **Any six from**: Avoid cross contamination/transfer from raw meat to cooked meat products by: foods touching/blood and juices dripping/transferring by hands, work surfaces/knives or equipment [1]; Good personal hygiene of workers – hand washing/clean protective overalls [1]; Good hygiene during cooking and serving – cover and cool all cooked meat as rapidly as possible/don't prepare too far in advance/no exposure to flies etc. [1]; Use red chopping boards [1]; Avoid incorrect storage, i.e. room temperature instead of below 8°C/not covering meat/store in bottom of refrigerator to avoid drip contamination [1]; Storage – use stock rotation/stick to use-by date [1]; Thaw meat thoroughly before cooking [1]; Do not undercook meat or bacteria will not be killed in centre/use a temperature probe to make sure that the correct temperature needed to kill bacteria has been reached [1]; Chilling – allow meat to cool before putting it into chill cabinets or the freezer/90 mins to chill below 8°C/use a blast chiller to cool quickly [1]; Reheat to the correct temperature for a long enough period of time (over 72°C) [1]; Hot holding – make sure hot meat products are kept at a hot enough holding temperature (63°C) [1]; Freezing – do not refreeze meat once it has been defrosted [1].

Page 168: Preparing and Cooking Food

1. **Any three from**: Store food in the correct place [1] because this reduces the chance of cross contamination/microorganism growth [1]; Store food at the correct temperature [1] because microorganisms require specific temperatures to grow, therefore, keeping them out of their temperature-growth zone slows bacteria growth [1]; Store food for the correct period of time [1] because storing food for longer than recommended increases the likelihood of microorganisms being present and growing in the food [1]; Defrost frozen products thoroughly [1] so that they can then be cooked to the correct temperature throughout – this ensures microorganisms in the middle of the food are killed [1]; Wear clean clothes when handling food [1] because this reduces the chances of contamination from clothing, e.g. dirt or pet hair [1]; Wash hands thoroughly and regularly, especially after using the toilet, handling rubbish or handling raw or different food products [1] because this reduces the chances of contamination from these sources [1]; Use clean equipment/clean the equipment thoroughly [1] because this reduces the chances of cross contamination

if the equipment was used for a different food item, e.g. raw and cooked meats [1]; Do not allow raw food to come into contact with cooked food [1] so that the chances of cross contamination are reduced [1]; Do not cough/sneeze over food or touch your nose/face when handling food [1] because this can spread bacteria/viruses present, leading to an increased risk of food poisoning [1]; Do not let animals or pests enter the food preparation area [1] because animals carry diseases and bacteria which can infect the food, and their hairs may also infect the food [1]; Cook food to the correct temperature [1] to ensure that microorganisms are killed to stop their growth [1]; Cook food for the correct amount of time [1] so that the food is cooked all the way through, with no cold spots – this ensures that all parts of the food have been heated to at least above the danger zone for microorganism growth [1]; Cool leftover food quickly [1] so that the time in which the food is in the danger zone is minimised, thereby reducing the chances of contamination and microorganism growth [1]; Reheat food only once [1] as food that has been cooked and cooled previously gives microorganisms more opportunity to grow [1]; Tie hair back/do not wear jewellery/false nails [1] to reduce the likelihood of these objects or bacteria from them falling into the food and contaminating it [1]. (**1 mark for each point made, with an additional 1 mark for explanation of each point – both must be given – up to a maximum of 6 marks.**)

2. **Any four from**: Avoid cross contamination [1]; Check use-by date [1]; Use appropriately colour-coded board/separate equipment [1]; Cook thoroughly/use a probe to check the temperature [1]; Once cooked serve immediately [1]; Use a clean knife/chopping board/work surface for preparation [1]; Wash hands before preparing/after handling [1]; Wear protective clothing/make sure hair is tied back [1].

3. **Any eight from**: Store foods according to their correct storage instructions [1]; Use correct cooking utensils for different foods [1]; Clean equipment and surfaces (with an antibacterial spray) [1]; Use within the best-before and use-by dates [1]; Clear up goods that have spilled [1]; Cook foods according to their cooking instructions [1]; Wash your hands [1]; Tie hair back [1]; Take off jewellery [1]; Remove nail varnish [1]; Wear a clean apron [1]; Take extra care with food preparation when ill [1]; Cover food [1]; Avoid coughing/sneezing over food [1]; Handle food as little as possible [1]; Cover cuts with waterproof dressing [1]; Avoid cross-contamination [1].

Page 169: Food Choices

1. Vegetarians eat fish. False [1]; Buddhists eat pork. False [1]; Sikhs don't eat beef. True [1]
2. c) [1]
3. c) [1]
4. **Any three from**: For health reasons [1]; For religious reasons [1]; Ethics – against cruelty to animals [1]; Ethics – against over-using the Earth's resources [1]; Don't like meat [1]; Born into a vegetarian family [1].
5. a) Lacto-vegetarians do not eat: meat/fish/eggs/non-vegetarian cheese/gelatine (from bones) [1]; which involves killing the animal [1]; but do eat animal products such as milk/cream/yogurt, etc. [1]
   b) Vegans do not eat anything sourced from animals [1]; such as: meat/fish/honey/gelatine/milk/milk products (cheese, butter, yogurt, cream) [1]; but do eat plant-based foods [1]

Page 170: British and International Cuisine

1. b) [1]
2. Wensleydale [1]
3. a) Cornish pasty [1]; hotpot [1]
   b) bouillabaisse [1]; Quiche Lorraine [1]
   c) tortilla [1]; paella [1]
   d) Peking duck [1]; chow mein [1]
   e) pakora [1]; dhal [1]
   f) minestrone [1]; focaccia [1]
4. **Answers depend on student's choice of country**

Page 171: Sensory Evaluation

1. smell [1]
2. a) [1]
3. a) Quiet area [1]; Invite people to taste [1]; Provide: water to cleanse palate [1]; Coded sample of food [1]; Clean eating implements, if needed [1]; Use a recording sheet [1].
   b) Testing two similar food products [1]; Three samples used but two the same [1]; All samples coded differently [1]; Aim is to try to identify the 'odd one out' [1].
   c) Name of product: any suitable name
   Samples to trial: any suitable answer, e.g. 15% fat minced beef [1]; 5% fat minced beef [1].

Page 172: Food Labelling

1. c) [1]
2. b) [1]
3. a) Guideline Daily Amount [1]
   b) As a percentage [1]; Per portion [1]; Per 100 g of food [1]
   c) **Any six from**: Weight or volume ('e'approximate weight) [1]; Ingredients list (from largest to smallest) [1]; Allergen information [1]; GM (Genetically Modified) ingredients [1]; Date mark and storage [1]; Cooking instructions – to ensure food is safe to eat [1]; Place of origin [1]; Name and address of manufacturer (in case of complaint) [1]; Lot or batch mark (for traceability) [1]; Nutritional information on pre-packaged foods [1].
4. b) [1]

Page 173: Factors Affecting Food Choice

1. b) [1]
2. c) [1]
3. a) Exercise improves cardiovascular and bone health [1]; Exercise helps to maintain a healthy weight [1]; Exercise improves self-confidence [1]; Exercise develops new social skills if taken with other people, e.g. team sports, aerobics classes [1].
   b) The Eatwell Guide [1]
   c) Variety of flavours [1]; Variety of textures [1]; Variety of colours [1].
   d) **Any four from**: The type of work done by people in the household will affect their appetites [1]; The number of hours worked by people in the household will affect their appetites [1]; The travelling (commuting) time of members of the household will affect the amount of time available for the preparation and consumption of food [1]; The pastimes of individuals will affect their appetite and the amount of time available for shopping, preparing and cooking food [1]; Whoever is in charge of the planning and cooking of food will influence the types of food bought and consumed [1]; The available income for buying food will affect the types of foods eaten [1]; If someone is vegetarian/vegan [1]; If someone has a food allergy/intolerance [1]

Page 174: Food and the Environment

1. At each stage of a product's lifecycle energy [1] is needed to process [1], transport [1] and dispose [1] of the product; carbon dioxide [1] is produced as a byproduct of energy use. The carbon footprint [1] is the calculation of the carbon dioxide produced throughout a product's life.
2. Deforestation [1] occurs when trees are cut down. Cleared land is used as pasture [1] for livestock [1] and plantations [1] of commodities and settlements. Deforested regions typically suffer soil erosion [1] and frequently degrade into wasteland [1].
3. Wise shopping and planning ahead reduces the amount of food bought in the first place [1]; FIFO (first-in first-out storage) reduces food wasted [1]; Only prepare the food you actually need, so nothing is needlessly thrown away [1]; Use food before it goes out of date, so that food does not have to be thrown away for safety reasons [1]; Use leftover food to make other dishes, thereby avoiding having to throw leftover food out [1]; Do home composting so that any food you have to throw out does not have to be transported to a landfill site [1].
4. **Any two suitable answers**, e.g.: Bubble and squeak [1]; Rissoles [1]; Soup [1]; Corned beef hash [1].

Page 175: Food Provenance and Production Methods

1. When traceability is fully available, trust is built between the retailer and the consumer [1]; Other criteria in which the consumer has an interest, such as ensuring the food is organic, vegetarian, specific allergen free, Kosher or Halal can be guaranteed via traceability [1]; This ensures that consumers can have confidence in the food they purchase [1]; Where there is a risk to public health, manufacturers may need to isolate sources, so traceability is practical [1].
2. Battery, **any two from**: Large numbers of hens kept in massive buildings designed to maximise growth [1]; Fed on high nutrient feeds over a short period of time [1]; Antibiotics and growth enhancers widely used [1].
   Free range: These hens have access to outdoor areas for part of their lives [1]; They do not live in cages [1].
   Barn, **any two from**: These hens live in an environment similar to intensively-reared animals but have access to natural light from windows [1]; They live in a lower density of animals per square metre [1]; They have access to environment enrichment such as fresh straw [1].
   Organic: hens are fed on products free from chemical or synthetic treatments that have relied on natural compost and manure for fertilisers [1]; Often kept out of doors with complete freedom [1].
3. b); c) [1]

Page 176: Sustainability of Food

1. CO_2/carbon dioxide [1]
2. **Any three from**: Increase crop diversity [1]; Improve soil organics by using animal waste [1]; Change the dependence on fossil fuels to transport foods [1]; Tackle deforestation issues [1]; Put in irrigation systems in drier areas [1]; Look at crop rotation to reduce soil erosion and the general health of crops [1]; Prevent soil erosion from winds, high rainfall and flooding [1].
3. a) **Any two from**: The Fairtrade logo means that the farmer in a developing country who produced

the goods gets a realistic income [1]; Investment in the local community takes place [1]; There are better working conditions for the producing farmer [1]; A fair price is paid for the goods [1]; Sustainable production methods are used [1].

b) Possible answers, **any two from:** Chocolate; Tea; Coffee; Bananas **(two answers needed for 1 mark)**.

4. The Red Tractor logo tells us that the food has been produced, processed and packed to the Red Tractor standards [1]; The flag on the Red Tractor logo shows the country of origin [1]; Red Tractor labelling assures good standards of food hygiene and safety [1]; Red Tractor labelling assures high standards of equipment used in production [1]; Red Tractor standards assure good standards of animal health and welfare [1]; Environmental issues are respected by Red Tractor suppliers [1]; Red Tractor standards ensure responsible use of pesticides [1]. Any product with the Red Tractor logo can be traced from farm to fork [1].

5. Livestock, especially cows, produce methane gas [1]; Methane gas is 20 times more harmful than CO_2 [1], with cows producing more Green House Gases (GHG) than the entire world's transport.

Page 177: Food Production

1. In homogenised milk, the milk is forced through tiny holes in a machine [1]; This breaks up the fat and disperses it, and it doesn't reform as a layer [1]
2. The cream may separate [1]
3. a) **Any three suitable answers, e.g. three from:** Red Leicester [1]; Cheddar [1]; Cheshire [1]; Lancashire [1]; Wensleydale [1]; Stilton [1]; Caerphilly [1].
   b) **Any three suitable answers, e.g. three from:** Brie [1]; Camembert [1]; Fromage frais [1]; Roquefort [1]; Saint Agur [1].
4. Possible answers, **any six from:** White [1]; Granary [1]; Wheatmeal [1]; Wholemeal [1]; Brown [1]; Spelt [1]; Self-raising [1]; Plain [1]; Strong [1]; 00 pasta flour [1].
5. Bread is made by mixing strong flour (which is high in gluten [1]) with liquid [1] and a raising agent such as yeast [1]. The yeast ferments with sugar [1] and warm water, and then when added to the flour and salt it makes a dough [1]. The dough is then worked by a process called kneading [1]. The dough is then allowed to rise further by standing covered in a warm environment. This is known as proving [1]. The bread dough is then shaped and finished with a glaze [1] before baking.
6. 00 flour [1]; Made from durum wheat [1]

Page 178: Food Processing

1. High temperature [1]; Low temperature [1]; Drying [1]; Chemical [1].
2. Method, **any one of:** Freezing [1]; Sugar [1]; Vinegar [1]; Oven-drying [1]. Shelf life, **corresponding one of:** Freezing – food is preserved for up to one year in temperatures between –18°C and –29°C [1]; Sugar – fruit is preserved with sugar, e.g. jam, for a couple of years [1]; Vinegar – vegetables can be preserved for up to two years by immersion in vinegar [1]; Oven drying – a warm oven can be used to dry foods slowly and they can then be stored in an airtight container for several months [1].
3. Low temperature does not affect nutritional value [1]
4. Meat/fish is 'cooked' by exposing it to heat from wood fires [1]; This gives it a distinctive smoky taste [1].
5. Fruit could be preserved in jars of alcohol – usually brandy [1]; Fruit could be added to sugar to make jams [1]; Vegetables could be pickled in vinegar to make pickles or chutneys and stored in jars [1]; Vegetables could be stored in jars in a brine(salt) solution [1].
6. They are pickled in vinegar with spices [1]
7. It is salted [1]; It can also be smoked [1]

Page 179-195 Practice Exam Paper 1

Section A

1.1	B	[1]
1.2	D	[1]
1.3	C	[1]
1.4	A	[1]
2.1	D	[1]
2.2	C	[1]
2.3	C	[1]
2.4	A	[1]
3.1	B	[1]
3.2	A	[1]
3.3	C	[1]
3.4	C	[1]
4.1	C	[1]
4.2	A	[1]
4.3	B	[1]
4.4	C	[1]
5.1	C	[1]
5.2	A	[1]
5.3	B	[1]
5.4	C	[1]

Section B

6.1 **Any four from:** When sugar is consumed it is broken down by the bacteria on the plaque found on teeth [1]; By this process, the sugar turns into acid [1]; This acid causes the tooth enamel to dissolve [1]; With damaged or non-existent enamel, the tooth is weakened [1]; As a result of weakened enamel, a cavity (hole) develops in the tooth [1]; The whole tooth can become damaged as a result of the cavity [1]; Pain may result because of the damaged tooth [1]; Fillings will have to be applied to the damaged tooth or a complete extraction will be necessary [1].

6.2 **Any four from:** Avoid sugary drinks – always check labels or use sugar apps to find out information about sugar content [1]; Choose water instead of a sugary drink [1]; Reduce consumption of high sugar foods, cakes, biscuits, chocolate and sweets [1]: Do not give sweets as a reward to children [1]; Avoid breakfast cereals coated with sugar, which are typically marketed to children [1]; Reduce the sugar content of foods when doing home baking – use natural fruits or dried fruits as an alternative to sugar [1]; Teach children about the dangers of overconsumption of sugar – help them to make the right decisions [1]; Parents should teach children by example – by not eating too much sugar themselves [1]; Check school food policy on sugar and on sugar in school dinners [1].

6.3 Named disease, **any two from:** Diabetes (Type 2) [1]; Obesity [1]; Coronary heart disease [1]. Causes – diabetes type 2: High blood sugar [1]; Lack of insulin/no insulin to regulate sugar levels [1]. Causes – obesity, **any two from:** Sugar is high in calories [1]; Over-consumption leads to excess fat forming [1]; Under skin and around internal organs [1]. Causes – coronary heart disease, **any two from:** High blood sugar level leads to diabetes type 2, which more than doubles the risk of developing coronary heart disease [1]; Lining of blood vessels become thick, which restricts blood flow [1]; Heart has to work harder to get oxygen around the body [1].

6.4 Two other carbohydrate groups: Starch [1]; Non-Starch Polysaccharide [1]. Examples – starch, **any two from:** Bread [1]; Pasta [1]; Potatoes [1]; Rice [1]; Breakfast cereals [1]. Examples – Non-Starch Polysaccharide, **any two from:** Wholegrain cereals [1]; Wholemeal bread [1]; Wholegrain breakfast cereals, e.g. bran flakes, Weetabix, shredded wheat, porridge oats [1]; Wholemeal pasta [1]; Wholemeal flour [1]; Any named fruit [1]; Any named vegetable [1]; Dried fruit [1]; Nuts [1]; Seeds [1]; Beans/peas/lentils [1]. **(2 marks for naming each type of carbohydrate group, 2 marks for two examples of each carbohydrate group, up to a maximum of 4 marks, 6 marks in total for the question)**

6.5 **Any six from:** Protein [1] – for growth, maintenance and repair of the body [1]; Some fat [1] – to provide concentrated sources of energy, and to aid brain function [1]; Carbohydrate (starch) [1] – for slow release energy [1]; Calcium [1] – for strong bones and teeth [1]; Iron [1] – for the formation of haemoglobin in red blood cells [1]. Fluoride [1] – to strengthen teeth [1]; Vitamin C [1] – to aid the absorption of iron and to build connective tissues [1]; Vitamin D [1] – to aid the absorption of calcium [1]; Reference to current nutritional guidelines [1] and the Eatwell Guide [1].

7.1 Herb – bay leaf [1]; Spice – peppercorns [1]

7.2 Milk contains lactose [1]; Lactose intolerance means that an individual is allergic to/cannot tolerate lactose in their diet [1]

7.3 Step 1: **Any three from**: hob heat [1]; Conduction though pan [1]; Pan base heat [1]; Liquid heat by convection [1].

7.4 **Any three from**: Starch in flour [1]; Swells [1]; Agitated beaten [1]; Smooth gelatinisation [1].

7.5 Choux pastry: Function – to give structure to the pastry [1]. Description – flour is used to thicken the pastry dough [1], the flour forms structure [1] Bread: Function – to provide gluten [1]. Description – **Any two from**: makes the dough [1], stretchy dough lets yeast work [1], sets on cooking [1]

8.1 **Any four from**: Pasteurised milk [1] – this extends shelf life [1]; Skimmed, pasteurised [1] – all or most of the cream is removed [1]; Semi-skimmed, pasteurised [1]; some of the cream is removed [1]; Ultra-Heat Treated (UHT), also known as long life milk [1] – has a shelf life of up to six months [1]; Sterilised, homogenised [1] – has a longer shelf life/has a slightly caramel flavour [1]; **Dried** [1]; Evaporating the water, leaving a fine powder [1]. **Canned; evaporated** [1] – Water evaporated off. Sweet and concentrated. Homogenised. Sealed in cans and sterilised [1]. **Condensed** [1] is evaporated milk that hasn't been sterilised; added sugar; very thick [1]. **(1 mark for the name of each type of milk and 1 mark for each reason.)**

8.2 **Any two from**: Butter [1] – cream is churned to make butter [2]; Cream [1] – the fat removed from milk is used [2]; Cheese [1] – this is milk in its solid form [2]; Yoghurt [1] – milk has a bacteria culture added to it [2].

9.1 **Any eight from**: Energy kcals comparison using data [1] better choice is sauce B [1]; Unsaturated fat comparison using data [1] better choice is B [1]; Saturated fat comparison using data [1] better choice is B [1]; Sugar comparison using data [1] better choice is B [1]; Comments about ingredients relating to healthy choices: Vegetables low in fat [1]; Vegetables low in kcals [1];

Vegetables should be included in a healthy diet [1]; Obese person should choose foods that are: Low in kcal [1]; Low in fat [1]; Include vegetables [1]. **(The eight relevant points chosen must avoid repetition.)**

10.1 Victoria sandwich not rising, **any two from**: Too much sugar, causing the gluten to be over-softened so that it collapses [1]; Too much raising agent, causing the gluten to overstretch and collapse [1]; Undercooking, caused by the wrong temperature or cooking time [1]; Opening the oven door before the gluten has set, so the heavy cold air makes it sink [1]. Lumpy cheese sauce, **any two from**: Liquid and starch not blended before cooking [1]; Insufficient stirring during the cooking [1]; Cheese added when sauce is cooled so does not melt [1]; Roux not cooked sufficiently [1]; Incorrect proportion of ingredients [1].

10.2 High risk foods are easily contaminated by bacteria [1] and so can cause food poisoning if not correctly stored at a temperature of 0°C-5°C and cooked thoroughly [1]. They have a short shelf life [1]. High risk foods include foods which aren't cooked before being eaten so, if contaminated, bacteria will not be destroyed e.g. cream, cooked meats, raw fish (sushi) [1]. Protein foods such as meat, milk, fish and eggs are high risk, as are cooked rice and lentils [1]. Other high risk foods are moist foods like gravy and soup and unpasteurised foods e.g. soft cheese made from unpasteurised milk [1].

Pages 196-212 Practice Exam Paper 2

Section A

1.1 B		[1]
1.2 D		[1]
1.3 B		[1]
1.4 A		[1]
2.1 C		[1]
2.2 B		[1]
2.3 A		[1]
2.4 B		[1]
3.1 A		[1]
3.2 D		[1]
3.3 B		[1]
3.4 D		[1]
4.1 B		[1]
4.2 C		[1]
4.3 D		[1]
4.4 C		[1]
5.1 C		[1]
5.2 A		[1]
5.3 C		[1]
5.4 D		[1]

Section B

6.1 **Any six from**: Three energy-giving macronutrients are identified: Carbohydrates are present within

breakfast in the form of starch and sugars [1]. Specifically, they will be provided from the bread and the hot chocolate drink [1]. Energy may be released in different ways depending upon the type of food. In this particular instance there will be a slow release of energy from the wholemeal bread [1]. Fat: Present within the breakfast in the form of saturated and unsaturated fats [1]. Specifically provided from the sausages, streaky bacon, whole milk and cooking fats [1]. There is a lot of fat in this breakfast that will deliver high energy content [1]. Protein: Protein is a secondary source of energy and is available in the egg, sausage and milk [1]. Any other relevant and correct response can be credited.

6.2 **Any 12 from**: High fat diets linked to high cholesterol [1], which attaches to sides of arteries, narrows them, restricts blood flow, blocks arteries [1], can lead to Coronary Heart Disease (CHD) [1]. Linked to High Blood Pressure, angina and stroke [1]. High fat diets are high energy possibly leading to obesity [1], which causes both physical and psychological problems [1]. In addition to a high fat diet, there are several factors that often work together to contribute to ill health and increased future health risks, these are some possible responses, but other relevant factors should be rewarded as appropriate: Lack of physical activity [1]; Psychological influences – for example, may use eating as a coping mechanism for dealing with emotional problems, such as family break-up, etc. [1]; Genetics – for example, family history of overweight people due to genetic reasons [1], family history of medical conditions [1]; Socio-economic issues – for example, low income backgrounds [1], lack of time, resources, knowledge, skills [1], reliance on fast foods [1], parents working and effect of each of these on food choices [1]. Also note that there are unhealthy dietary options which do not reflect current government dietary guidelines, such as the Eatwell Guide [1].

6.3 **Any three from**: Energy linked to Kcal [2]; Warmth/insulation linked to body fat [2]; Protection of internal organs – kidneys [2]; Source of fat-soluble vitamins A, D, E and K linked to relevant food sources [2]; Formation of cell membranes – maintenance and good health [2].

6.4 Saturated fat, description of make-up, **any two from**: Contains the maximum amounts of hydrogen [1]; Molecule made up of single bonds (diagram could be drawn) [1]; Solid fat [1]; Solid animal fat [1].

Examples, **any two from**: Butter [1]; Lard [1]; Ghee [1]; Dripping [1]; Suet [1]; Cream [1]; Coconut oil [1]. Unsaturated fat, description of make-up, **any two from**: Able to accept more hydrogen [1]; More than one double bond in the molecule (diagram could be drawn) [1]; Vegetable source – oils [1].

Examples, **any two from**: Vegetable oil [1]; Corn oil [1]; Olive oil [1]; Rape seed oil [1]; Sunflower oil [1]; Groundnut oil [1]; Sesame oil [1]; Some fish oils [1]; Commercial fats, e.g. Flora products [1].

7.1 Flavouring – lemon zest [1]; Spice – cinnamon [1]

7.2 Butter is a saturated fat and not a healthy choice [2]

7.3 Oven is pre-heated [1] convection currents [1] baking tray heats shortbread by conduction [1].

7.4 **Any three from**: Fat melts [1]; Fat is then absorbed by flour [1]; Surface sugars caramelise [1]; Shortbread crisps [1].

7.5 Pastry tarts: Function – Shortening [1] Description – Fat coats the flour grains, preventing gluten development [1] so the cooked texture is short and crumbly [1]. Sponge cake: Function – Aeration [1] Description – When fat and sugar are creamed air is trapped [1] so the cooked texture is light and fluffy [1].

8.1 **Any six from**: Farmers' markets contribute to thriving local economies [1]; Sustainable livelihoods [1]; Protects the diversity of both plants and animals [1]; Welfare of farmed and wild species [1]; Farmers' markets avoid dependence on fossil fuels to transport foods around the world [1]; Farmers' markets sell quality locally grown products that support local businesses and farmers [1]; Farmers' markets provide fresh produce [1]; Farmers' markets avoid the need for extra packaging or the use of preservation techniques [1].

8.2 **Any three from**: Battery (intensive farmed) [1]; large numbers of hens kept in massive buildings/designed to maximise egg production/on high nutrient feeds/short period of growth time/antibiotics/growth enhancers/fertilisers and pesticides being widely used/poor quality eggs [1]. Barn [1]; environment similar to intensively-reared animals/have access to natural light from windows/live in a lower density of animals per square metre/better quality eggs than intensive farmed [1]. Free range [1]; allows animals or poultry access to outdoor areas for part of lives/behave as they would in nature, digging and foraging/hens that are free-range produce eggs that are more nutritious and tasty [1]. Organic [1]; hens reared naturally without help from any chemical or synthetic treatments/rely on natural composts and manure for fertilisers for growing their feed/are GM free/no proof that organic food is more nutritious/buying organic food is a lifestyle choice/organic and free-range farming more ethical/lower negative environmental impact [1].

8.3 **Any two from**: Hens' eggs must carry a stamp with a number indicating whether they have been produced in an organic, free-range, barn or cage system [1]; Egg boxes must clearly state: 'eggs from caged hens', 'barn eggs' or 'free range' [1]; Eggs must carry the lion mark [1]

9.1 **Any eight from**: Both [1] of these sausages are gluten- free [1] so can be eaten by a coeliac because they will not contain gluten [1]; Saturated fat comparison using data [1], better choice is sausage A [1]; Energy kcals comparison using data [1], better choice is sausage A [1]; Dietary fibre comparison using data [1], better choice is sausage A [1]; Coeliacs would look for the gluten-free label or symbol [1]; Comments should be included about ingredients relating to healthy choices: Vegetables are low in kcals/vegetables should be included in a healthy diet [1]; People should choose foods that are: Low in kcal [1], Low in fat [1]; Contain protein [1]; Contain dietary fibre [1]; Are low in salt [1]; Are low in total fat [1]. **(The eight relevant points chosen must avoid repetition.)**

10.1 Bread not rising, **any two from**: Liquid too warm it kills the yeast [1]; Room is too cold during proving [1]; Insufficient proving time [1]; Too much salt added, which kills the yeast [1]; Insufficient liquid used [1]. Quiche not setting, **any two from**: Not enough egg mixture to set the quiche [1]; Too much liquid (milk) [1]; Oven temperature too low [1]; Insufficient cooking time [1].

10.2 **Any six from**: Follow storage instructions on label to prevent food spoilage [1]; Chilled foods should be stored in a refrigerator between 0°–5 °C as this slows down bacterial growth [1]; Use a fridge thermometer to ensure the fridge is kept at a safe temperature, which should be between 0° and 5 °C [1]; Cooked foods should be stored above raw foods to prevent cross contamination [1]; Cover foods in the refrigerator to prevent cross contamination [1]; Check foods frequently for signs of decay and that they are within their use-by dates, as bacterial growth can lead to food poisoning [1]; Keep refrigerators clean to prevent microbial growth [1]; Do not overload the fridge as air needs to circulate to keep the food cool [1]; Do not put warm foods in a fridge as this will raise its temperature [1]; Store frozen foods in the freezer below –18°C as this prevents food thawing out and bacteria becoming active [1]; Do not re-freeze food once it has defrosted as bacterial multiplication may have taken place as the food warmed up [1]; Rotate stock/check use-by dates to ensure the oldest food is used first – this prevents dry foods becoming rancid or infested by pests [1]; Store non-perishable foods in cool dry conditions, to keep them fresh [1].

Notes

Notes

Notes

Notes

ACKNOWLEDGEMENTS

The author and publisher are grateful to the copyright
holders for permission to use quoted materials and images.

p.6-7 Bridge hold and claw grip. This resource was developed for the DfE
Licence to Cook programme 2007-2011. © Crown copyright 2007.
http://www.nationalarchives.gov.uk/doc/open-government-licence/version/3/
p.12 Peeling food, grating food, stirring food. This resource was developed for
the DfE Licence to Cook programme 2007-2011. © Crown copyright 2007.
http://www.nationalarchives.gov.uk/doc/open-government-licence/version/3/
p.28 The Eatwell Guide. © Crown Copyright 2016. Contains public sector
information licensed under the Open Government Licence v3.0.
http://www.nationalarchives.gov.uk/doc/open-government-licence/version/3/
p.83 © Food Standards Agency
p.99 Libby Welch / Alamy Stock Photo
p.95 Carolyn Jenkins / Alamy Stock Photo
All other images © Shutterstock.com

Every effort has been made to trace copyright holders and obtain their
permission for the use of copyright material. The author and publisher will
gladly receive information enabling them to rectify any error or omission in
subsequent editions. All facts are correct at time of going to press.

Published by Collins
An imprint of HarperCollinsPublishers Ltd
1 London Bridge Street
London SE1 9GF

© HarperCollinsPublishers Limited 2020
This edition published 2020

ISBN 9780008166342

First published 2017

10 9 8 7

British Library Cataloguing in Publication Data.

A CIP record of this book is available from the British Library.

Authored by: Kath Callaghan, Fiona Balding, Barbara Monks,
Barbara Rathmill and Suzanne Gray with Louise T. Davies
Contributor: Sarah Middleton
Commissioning Editors: Fiona Burns and Katherine Wilkinson
Project Editor: Katie Galloway
Editorial: Katie Galloway and Jill Laidlaw
Cover Design: Kevin Robbins and Sarah Duxbury
Inside Concept Design: Sarah Duxbury and Paul Oates
Text Design and Layout: Jouve India Private Limited
Production: Natalia Rebow
Printed in the UK, by Martins The Printers
Published in association with the Food Teachers Centre.

MIX
Paper from
responsible source
FSC **FSC™ C007454**
www.fsc.org

This book is produced from independently
certified FSC™ paper to ensure responsible
forest management.

For more information visit:
www.harpercollins.co.uk/green